Mark Thompson is one of the presenters on the BBC One award-nominated show *Stargazing Live* and the regular astronomer on ITV's *The Alan Titchmarsh Show* and *This Morning* and is a regular on Radio Five Live.

When not gracing our television screens he is most likely writing about the sky, looking at it or flying through it (as a qualified pilot, not a human cannonball).

He writes for a number of websites including Discovery News, along with a variety of other publications. His recent Stargazer tour has been a hit, receiving a great review from the Edinburgh Festival.

Born in Norfolk, Mark has had a fascination with all things in the sky ever since he was a small boy and has served as a member of the Council of the Royal Astronomical Society.

www.markthompsonastronomy.com

D0727687

79 713 013 6

A Down to Earth Guide to the Cosmos

Mark Thompson

CORGI BOOKS

TRANSWORLD PUBLISHERS
61–63 Uxbridge Road, London W5 5SA
www.transworldbooks.co.uk

Transworld is part of the Penguin Random House group of companies
whose addresses can be found at global.penguinrandomhouse.com

Penguin
Random House
UK

First published in Great Britain in 2013 by Bantam Press
an imprint of Transworld Publishers
Corgi edition published 2015

A CIP catalogue record for this book
is available from the British Library.

ISBN
9780552170390

Typeset in 11/15pt Berkely Book by Falcon Oast Graphic Art Ltd.
Printed and bound by CPI Group (UK) Ltd, Croydon, CR0 4YY.

Penguin Random House is committed to a sustainable
future for our business, our readers and our planet. This book
is made from Forest Stewardship Council® certified paper.

MIX
Paper from
responsible sources
FSC® C016897

1 3 5 7 9 10 8 6 4 2

To Karen, Phoebe and Riley, the brightest stars in my sky

Contents

Introduction

NEARLY EVERY HUMAN on the planet has seen it, many are captivated by it and a few understand it, at least in part. I speak of course about the night sky, one of the natural world's most inspiring and beautiful sights and one that has had a special place in our hearts and minds for centuries. To the beginner, a star-filled sky can seem almost as mysterious and unfathomable as it was to our ancestors hundreds of years ago, but with this book as a guide the awesome nature of the cosmos can be enjoyed by us all.

I've been an amateur astronomer for nearly three decades now and I've lost count of the number of times I've met people who have an interest in the sky but just have not found the time to go outside and look up. When I finally encourage them to come out under the stars, they are often amazed that they can see so much without a telescope or binoculars. I can point out neighbouring planets, galaxies so far away that we see them as they were millions of years ago, glittering star clusters and ghostly glowing clouds forming slowly into stars. The effect is predictable: it never fails to impress and I have seen it time and time again. Just realizing that so much is on view can ignite a spark that has lain dormant for years and can start them on a fascinating journey around our Universe, a journey which we shall now embark upon together.

The real beauty of astronomy is that it is free and open to anyone, no matter what your age or ability. It does not have to cost you a penny to get started; all you need to do is open your eyes and look up to marvel at the cosmos. On your first night under a clear sky, you might see the odd shooting star or maybe a few satellites as they orbit the Earth, but apart from that you'll see a bewildering, perhaps daunting, mass of stars. Once you get out under that canopy, and with my help, you will soon learn your way around and be picking out celestial wonders.

Learning about the night sky requires two distinct areas of focus: the physical nature of the processes and objects among the stars, and an ability to navigate around the sky to find them. This book combines these two areas in a unique and engaging way, accompanying you in just twelve chapters on the same journey that has taken mankind thousands of years. The easiest way to start the journey is by looking at the stars. Getting people to engage with the sky above them is an approach I have used very successfully on the BBC's top-rated programme *The One Show* and again with my pupil Jonathan Ross in *Stargazing LIVE*. Through the course of this book I will teach you the fundamental concepts of modern astronomy and, by using the night sky to illustrate them, help you learn and remember your way around the stars.

Most people think you need to rush out and buy a telescope to start 'doing' astronomy. That could not be further from the truth. In fact, a telescope will more than likely hinder the development of some key skills that it is more important you should acquire first, so take a look at my Six Top Stargazing Tips to set you on the right track.

STEP 1 Start reading this book

You've made a great start by getting hold of this book. In its pages you will find information about the Universe and some fantastic star guides to help you start learning your way around the sky. You've already started on your journey to become an astronomer by reading this so carry on and let me be your guide around the cosmos.

STEP 2 Buy a red torch

As an astronomer you need your eyes to become adjusted to seeing in the dark to maximize what you can observe in the night sky. There are two changes that occur when you move from an illuminated environment, such as your lounge, to the dark of the night. First your pupil will fully open up, or dilate, allowing as much light in as possible. This only takes a few seconds but is followed by a chemical change inside your eye, optimizing it for night-time vision. It takes about forty minutes away from light for your eyes to become fully 'dark adapted' and exposure to bright light will immediately spoil this, but there are times when a little light is needed (to read the star guides in this book, for example). Red light is much less damaging to your eyes' ability to adapt to the dark and dim red light is ideal. I've heard some people recommend a bicycle rear light for astronomy but that can still be too bright, although it will be better than white light. Instead, you are much better off going to a specialist astronomy supplier, who will be able to provide you with a purpose-designed torch.

STEP 3 Subscribe to astronomy magazines

There is no better way to keep up to date with what is going on in the world of astronomy than subscribing to a specialist magazine. Most countries have at least one or two good ones dedicated to astronomy. Between the covers you will find news, equipment reviews, classified adverts and even monthly sky charts (but be sure to buy a magazine for your own country or these charts may not be relevant to where you live).

STEP 4 Find and join your local astronomical society

A great way to enhance your new hobby is to seek out your local astronomical club or society. These are great places to go for advice and eventually you will find that astronomy with your new-found friends makes observing sessions much more enjoyable. When you eventually decide to buy your first telescope (see step 6) it will be a great place to get to see the different types in action. Getting involved at your local club may lead to helping run public outreach events, talking to newcomers about your experiences and perhaps even lecturing or helping to run the society.

STEP 5 Get outside and start enjoying the wonders of the cosmos

After you have started on this book, got your red torch, maybe begun reading magazines and perhaps made contact with your local astronomical society, the next thing you need to do is the most exciting bit . . . and that is to get outside and start learning your way around the night sky. You will soon be amazed at what you can see; those bright stars which were not there before or do not appear in any sky guide are probably planets, and on your first night under the stars you may spot satellites, meteors or the odd passing aircraft. Time spent now familiarizing yourself with the sky will make your future enjoyment of the Universe much easier. This book will help start you off by identifying the brighter stars in the sky and showing you how to 'star-hop' from these easy-to-find stars to other, fainter ones. Soon you will be identifying constellations and before you know it you will recognize them without even looking at this book.

STEP 6 Consider future equipment purchases

Eventually you will want to make that all-important purchase and buy yourself a pair of binoculars or maybe even a telescope. The only words of advice here are to take your time and make sure you spend wisely. This book has a chapter dedicated to equipment and members of your local society will have a wealth of knowledge to help you with your purchase. It is worth starting with binoculars though and moving on to a telescope a little later.

A few words next on how this book works so you can start getting the most out of it straight away. The twelve chapters of the book represent the twelve months of the year and each chapter will tackle an astronomical topic; for example, Chapter 1 looks at our changing view of the universe, and will occasionally draw on examples from the night sky to illustrate the point.

Following on from the astronomical topic are the two 'Quick Sky Guides' to the stars for that month, one for the stars in the northern hemisphere of the sky and the other for those in the sky's southern hemisphere. To use them you will need to locate an imaginary line called the celestial equator (which is just an extension of our own equator out into the sky), which we'll use as the starting point to navigate together around the heavens.

I shall explain in the section that follows just how to find the celestial equator from where you are and give you a few more tips on using the sky guides in each chapter. Take time reading through this section to ensure you get the most out of the book.

. . . and that is it. Follow these simple steps and get started on the first chapter to see how astronomers over the centuries have slowly unlocked the secrets of the Universe. Before you know it you will be hunting down moons around other worlds, stellar nurseries or those elusive galaxies, and I promise you now, you will not be disappointed.

How to Use the Quick Sky Guides

O F ALL THE THINGS that I have learnt when trying to help people find their way around the night sky the most useful is to have something like a story or interesting fact to hook on to. It is just how the mind works: give a piece of information something to cling on to and you will have a much better chance of committing it to your long-term memory. Doing this can not only bring the sky to life but also help you to retain what you have learnt about the objects in it and, more importantly, remember where they are.

My 'Quick Sky Guides' cover first the sky in the northern hemisphere and then the sky in the southern hemisphere. It is important to add that there is a subtle difference between the northern and southern hemispheres of the sky, and the sky as it is visible from the northern and southern hemispheres of the Earth.

To make that a little clearer: if you stood at the North Pole, then you would only see the northern hemisphere of the sky; and from the South Pole only the southern hemisphere of the sky would be visible to you. If you were to observe the sky from a point somewhere along the equator, then one half of what you could see would be a part of the northern hemisphere sky, and the other half part of the southern hemisphere sky.

This means that, unless you live at either of the poles, you will be able to see part of both hemispheres of the sky, and how much of each is visible depends on your location. Therefore, to get the most out of them, you will need to look at parts of both sky guides.

The key to making this all work is in identifying an imaginary line called the celestial equator, which is 'visible' from anywhere on Earth, and it is here that each hemisphere starts before heading either north or south. The great thing is that its position in the sky relative to where you are will not change, at least not for a few thousand years, so once you know roughly where it is your exploration of the night sky can begin.

A tiny piece of mathematics is needed to find the celestial equator from your location. The first thing you need to do is find your latitude, which is a measure of how far north or south of the equator you are. With the vast array of mobile devices on the market which use various technologies to pinpoint your location on Earth, it is a simple matter of reading off your latitude. It will probably be given in degrees, minutes and seconds, but it is just the degrees you need. These are degrees like the degrees of a circle you learnt about in school, and you do not need to be that accurate. If you do not have access to this kind of device, get hold of a map and use its grid reference to find your latitude – again, pin-point accuracy is not crucial. Either way, you will have a number between zero and 90; if you live near the equator then it will be towards the lower end, if you live nearer the poles, then it will be towards the higher end.

Now here is the maths bit: subtract your number from 90 and that will give you the maximum height above the horizon of the celestial equator from where you are. All you need to know now is how to translate that to the sky and you can get on with the book. There is a really useful scale using your hand which helps you to estimate distances in the sky. A hand at arm's length with fingers spread measures about 25 degrees from thumb to little finger, and a clenched fist equals around 10 degrees. Smaller measurements can be gained from the width of the three middle fingers equivalent to 5 degrees and one finger's width equal to about 1 degree. To make that a bit clearer, see the illustrations below.

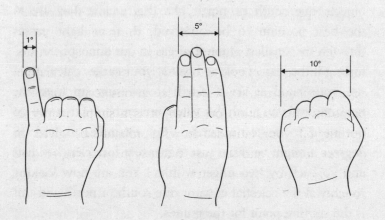

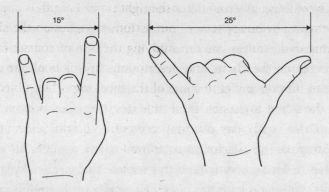

Angular Hand Measurements (not to scale)

In the northern hemisphere you need to turn and face south (if you are not sure which direction that is, when you look towards where the Sun sets your left shoulder is pointing south), or face north when in the southern hemisphere (the direction your right shoulder points when you look towards the position of the setting Sun). Looking at the objects due south or north like this means they are in the best position to be observed, as their light passes through the smallest amount of gas in our atmosphere, distorting it the least. Looking back at your earlier calculation, use your hand at arm's length to measure up from the horizon until you reach this value, remembering that you do not need to be too accurate. (For example, to find 38 degrees it might be easier just to measure four clenched fists and go back by two finger widths.) You are now looking roughly at the celestial equator due south or north and this is the starting point for the guides.

Now to get you looking at the right time of night. It might seem an obvious statement but astronomy relies on the sky being dark, unless you are studying the Sun of course. For this reason the charts and descriptions in this book are all set to introduce you to some of the most important objects in the sky at midnight, local time (the time shown by your watch). On any day during the month, go outside around midnight – do not forget to allow for any daylight saving scheme used in your country – and the stars will be in roughly the right position. As the months progress and you follow these guides you will notice that the stars move a little further to the west from month to month, giving a slightly different view of the sky; Arcturus, for example, was in the east at midnight in April and will be a little more to the west in June. This is because the Earth actually takes 23 hours, 56 minutes and 4 seconds to rotate once on its axis and we call this the sidereal day, although our watches measure a day as 24 hours. The effect of this is that the sky slowly gets out of synchronization with our watches.

Start with the section of the hemisphere you live in, and once finished switch to the other section and follow it until you meet your horizon. One more important thing to remember: if you live in the northern hemisphere, you must reverse the east/west references when using the southern sky guides and, likewise, southern hemisphere observers must reverse the east/west from the northern sky guides. You will also see references to magnitude in the guides as an indication of brightness. Astronomers use this scale to measure brightness, where a lower number is brighter than a higher number; for example, magnitude 6 is fainter than magnitude 2 and magnitude 2 is fainter than magnitude −1. To give you a gauge, magnitude 6 is the faintest star the average human can detect on a dark moonless night.

Now, before we get started, the following diagrams highlight the most important constellations in each hemisphere. These are not all visible at any one time, but will come into focus as we progress through each monthly guide.

Highlights of the Northern Hemisphere

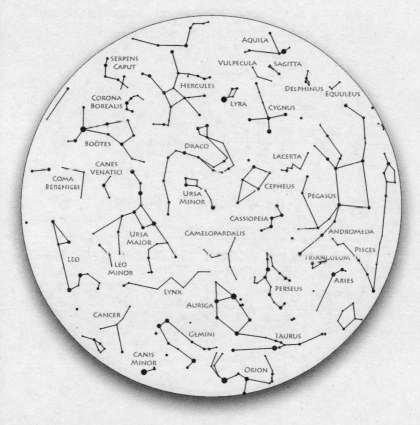

Highlights of the Southern Hemisphere

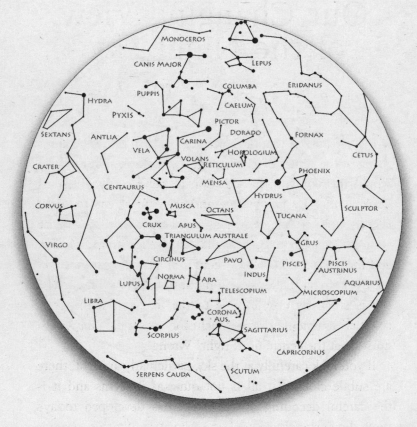

ONE

Our Changing View of the Night Sky

I T IS NO SURPRISE that our ancestors used to think the Earth was at the centre of everything. If you think about it, why would they, or you, have reason to believe any different? Look at the sky in daytime and the Sun seems to move silently around the Earth, rising in the east and setting in the west, taking a day to complete the trip. Look at the night sky and the stars and Moon all seem to travel in a path around us. Even if you make the assumption that everything in the sky is stuck by cosmic glue to the inside of some great celestial sphere then the conclusion is simple: the Earth must be at the centre of it, right? Wrong!

If you look carefully at the sky, particularly at night, there are subtle clues that all is not quite as it seems and it is the careful decoding of these that has developed today's three-dimensional view of the Universe. In this chapter we are going to take a look at the incredible story of our changing view of the cosmos and at some of the insightful observations of philosophers and astronomers through the centuries.

While it is fair to say not a lot of thought went into the Earth's place in the cosmos during the early days, there was still an awareness of the motions of the objects in the sky. We know that prehistoric hunters used the sky to mark the passage of time because ancient animal bones have been found with marks thought to represent the changing phases of the Moon. And who can forget the mysterious Stonehenge, which is thought to have been completed as far back as 1600 BC and is believed by many to be an ancient astronomical observatory? It is clear that our ancestors were aware of the sky, but they seemed to be passive observers for many centuries.

It is surprising that we can go all the way back to around 450 BC, to the time of the Greek philosopher Anaxagoras, and find that scientific reasoning was already slowly starting to sow the seeds for our current view of the Universe. Anaxagoras suggested that the Sun was a flaming ball of metal 'even larger than Peloponnesus', the large peninsula of southern Greece, and that we saw the Moon because it simply reflected light from the Sun. But the real turning point came around 200 BC, with one of the most insightful observations of all time, not only the realization that the Earth was round but more incredibly the measurement of its circumference.

The Earth is as we know spherical in shape, or more accurately an oblate spheroid (a sphere very slightly squashed at the poles), and we have been brought up with that knowledge, reinforced in modern times by stunning pictures from space. The evidence, though, is there in the sky for anyone to see and amazingly the information is also there for anyone with a basic knowledge of geometry to work it out for themselves.

Eratosthenes was another Greek philosopher who seems to have been annoyingly good at many things from mathematics to poetry and astronomy to athletics! He had heard that at the summer solstice in a town called Syene (now Aswan) in Egypt, the Sun shone directly down to the bottom of a deep well at noon. This meant the Sun was directly overhead, but at the same time in Alexandria the Sun was casting shadows, so could not have been overhead. You may have noticed yourself on your holidays how the Sun climbs higher or lower in the sky depending on where you travel to on the Earth. Eratosthenes realized this but took it a step further.

Knowing that Syene and Alexandria were in a direct north–south line, he set about using this information to work out the circumference of the Earth. The story goes that he actually got someone to walk between the two towns to measure the distance and found they were separated by 5000 *stadia*. Now, unfortunately, the exact measure of one stadion is unknown but a best guess from historical texts is that 5000 *stadia* equate to about 925km.

Eratosthenes realized that because he knew the distance between the two locations and the apparent shift in position of the Sun in the sky between them both, he could estimate the Earth's circumference. Plugging the numbers into some reasonably simple formulae he came up with a figure of approximately 250,000 *stadia*, or 46,250km, which is remarkably close to today's figure of 40,075km. This difference in the distances is largely due to experimental error and the inaccuracy of the measurements used but, even so, the result is pretty impressive without a calculator and for around 200 BC. However, though his calculation was accepted by at least some of his peers at the time, it took many years, even centuries, before this view of the shape and size of the Earth became widely accepted.

The next great leap in understanding our place in the Universe came from the movement of objects within it. For many centuries it was known that there was a special group of 'stars' which seemed to move or wander around the sky; in fact, their modern name, planets, comes from the Greek word *planetes*, via the Latin *planeta*, which means 'wanderer'. In its earliest form, the model of the Solar System known as the geocentric model had the Earth at the centre with the Sun, Moon, planets and stars fixed upon great crystalline spheres all revolving around the Earth. Beyond the last sphere of the stars was the realm of the gods. While everyday experience suggests this might be accurate, careful study of the sky reveals that the planets seem to move at different speeds and sometimes even backwards. The geocentric model did not quite account for this movement so an alternative explanation was needed.

The backward, or retrograde, motion was later explained in a modification to the geocentric model by Claudius Ptolemy in the second century AD, and, although complex, it did seem to make sense of this odd planetary motion. Ptolemy suggested that each planet was not orbiting the Earth but instead orbited a point (the epicycle), which in turn travelled along a path (the deferent) around the Earth. This meant the planets' orbits described a rather strange corkscrew pattern through space. In time, the model was modified even more, becoming increasingly complex to account for the observed planetary movements.

The model changed around the sixteenth century when the Polish astronomer Nicolaus Copernicus made the rather bold yet unpopular suggestion that the Earth was not at the centre of everything. Copernicus was also a Catholic cleric and making statements to dethrone the Earth from its central position was not popular with the church. His idea was not new though, as it had been first suggested as far back as the third century BC by Aristarchus, but it was not until careful and precise observations of the planets by Copernicus that evidence started to mount. He realized that it was possible to do away with the complex patterns of epicycles and deferents if the Sun rather than the Earth was at the centre of the Solar System, and if the Earth orbited the Sun like all the other planets.

The key reasoning behind Copernicus' theory was that the strange movement of the planets was just an apparent and not an actual one. In the same way that a fast car overtakes a slower car, the slower one 'seems' to move backwards although in reality it is still going forwards. Unfortunately the new Sun-centred view of the Solar System still required a number of epicycles and deferents but nowhere near as many as were needed in the Earth-centred model.

One of the main issues holding back scientific reasoning and the development of new ideas in these early years was the supremacy of religion. It was considered that all things in the sky from the Moon to the distant stars were divinely created so were nothing short of perfect. The only shape deemed to be perfect was the circle so all things in the sky should be circular and, if they moved, they should move in circular orbits. The reliance on keeping everything circular proved to be a sticking point for some time, and had philosophers and scientists overcome this they might have reached the right conclusion much earlier instead of getting distracted by epicycles and deferents.

The turning point was finally reached by Johannes Kepler, who for a good part of his life was assistant to the renowned Danish astronomer and nobleman Tycho Brahe. In 1572 Brahe observed a 'new star' in the constellation of Cassiopeia and noticed its gradually fading light. He also discovered that, due to a lack of parallax (which is an apparent shift in a star's position in relation to the Earth's movement in space), it must be at a very great distance from the Earth and therefore be on the outer sphere of the stars and well beyond the planets. This discovery did not sit well with the church as the heavens were supposed to be perfect and unchanging.

Brahe was a prolific observer and left it to Kepler to analyse his countless observations of the planet Mars. Unfortunately for Kepler, Brahe was incredibly protective over his observations so he only let Kepler look at them when absolutely necessary and certainly did not let him copy the records so he could spend more time over them. It was the death of Brahe in 1601 that proved to be a catalyst for Kepler and his work. He was named as Brahe's successor as Imperial Mathematician to the Holy Roman Emperor Rudolph II. In his new role, Kepler was charged with providing astrological data to the Emperor in an era when there was no differentiation between astrology and astronomy: it was merely a study of the sky. Kepler straddled what we might now consider two modern disciplines in both providing guidance to the Emperor and studying the physical properties of the sky. In his quest to understand the nature of orbits, he hit on an idea which was a precursor to the property we know as gravity. He suggested that the Sun, which like Copernicus he accepted to be at the centre of the system, had a force that was making the planets move, and even suggested it would be weaker at greater distances. This would make the planets further from the Sun move more slowly.

The discovery for which Kepler is best known came from his continued study of Brahe's observations, which he now had complete access to. These accurate observations of the position of the red planet did not fit with the prediction of the Copernican model, and to account for the discrepancy an ever-increasing level of complexity with more epicycles and deferents had to be employed. He even tried changing the nature of the supposed spherical orbits for an ovoid, or egg-shaped, system but this too proved to be far too complex. After almost fifty attempts at tweaking the theory to match observation he abandoned the use of the ovoid and tried elliptical orbits along a plane. Finally he had hit on a solution that worked and in 1609 was able to articulate this in his three laws of planetary motion:

1. *Planets move around the Sun in ellipses, with the Sun at one focus.* Essentially this simply says that the distance between a planet and the Sun varies.

2. *The line connecting the Sun to a planet sweeps out equal areas in equal times.* When the planet is closer to the Sun it moves faster than when it is further away.

3. *The square of the orbital period of a planet is proportional to the cube of its mean distance from the Sun.* Sounds scary but all it means is that you can work out how far away a planet is from the Sun if you know how long it takes to complete an orbit.

Not only did Kepler's laws of planetary motion beautifully explain and predict planetary orbits in the Solar System, but when the moons around the outer planets were discovered, the laws held firm to explain their motion around the parent body. Along with Newton's law of gravity the science is still being used today to send spacecraft to the planets and to understand the movements of planets around other stars.

In 1608, the year before Kepler published his laws of planetary motion, another pivotal discovery had been made.

Quite how the Dutch spectacle-maker Hans Lippershey came across the idea is unknown, but to him is attributed the invention of a device 'for seeing things far away as if they were nearby', which we now call the telescope. Just a year later the Italian astronomer and philosopher Galileo Galilei heard about the device and going on a rather sketchy explanation managed to re-create one for himself with 3x magnification. He later made some enhancements to the design and managed to achieve a higher magnification of 30x.

Galileo is most famous for being the first person to turn one of these newly invented telescopes to the sky and in doing so making some incredible discoveries. He first used his telescope to look at the planet Jupiter in January 1610 and found what he described as 'three fixed stars invisible by their smallness' that lie along a line with the planet. He noted that they moved position from night to night and on one occasion one of them even disappeared. He concluded that these strange objects must actually be in orbit around the giant planet and discovered a fourth just a few days later. He had found four of Jupiter's moons: Io, Europa, Ganymede and Callisto. This discovery had far-reaching implications though, as it was the first piece of evidence that bodies orbited around something other than the Earth, it supported the idea that the Earth was not really that special in our Solar System after all.

Later the same year Galileo studied the planet Venus and noticed that he could track a full series of phases just like the phases of the Moon. He realized that this too was evidence of the Copernican model of the Solar System with the Sun at the centre since this simply could not happen in the Earth-centred system. Galileo's discoveries were not only ground-breaking but instrumental in turning the tide from religious belief governing astronomy to science and observation.

Up until now, we've seen how it was discovered that the Earth was a sphere, that the Sun was at the heart of our planetary system and that all planets moved in elliptical orbits. The invention of the telescope and subsequent observations set strong foundations for our modern view of the Solar System but still the distances to the stars and the nature of deep space proved to be a mystery.

The next leap forward came almost 230 years later – after the German mathematician and astronomer Friedrich Bessel became the first to measure the parallax of a star. He studied the star 61 Cygni in the constellation of Cygnus and found it seemed to shift by 0.314 of an arc-second (1/5400th the size of the full moon), which was caused by the Earth's changing position in its yearly journey around the Sun. When this tiny measurement is plugged into a special formula it reveals that the star is at a distance for this shift of 10.4 light years. By the end of the 1800s around sixty stars had been identified as displaying a measurable shift in their position and for the first time in history we had a real understanding of interstellar distance, at least in the nearby Universe. But for the rest of the stars, it remained a mystery.

Telescopic studies also revealed that the strange hazy

band of light from the Milky Way arching across the sky was actually millions of distant glittering stars. Quite why there was a concentration of stars in a band across the sky was a subject for much discussion among astronomers. The answer eluded them for nearly two centuries after Galileo's initial observations but the development of even larger telescopes proved to be crucial in finding the explanation.

William Herschel was a musician who pursued astronomy in his spare time, much like today's amateur astronomers. His ambitions, though, were big, matched only by his passion for the night sky, so he built his own large telescopes and turned them on the sky to discover new worlds like the planet Uranus. He also spent his time counting stars, a time-consuming and tedious activity, but on counting their numbers in different parts of the sky, he believed he could find the area of highest concentration, which would be the direction of the galactic centre. His search found no such area so he concluded the Sun and its system of planets must sit at the centre of a vast disc-like structure.

It took another hundred years before an astronomer working at the Mount Wilson Observatory in the United States came up with a more accurate and reliable method for estimating the size and scale of the Milky Way. Harlow Shapley was using a huge 1.5m reflecting telescope to study a type of object called a globular cluster. As their name suggests, these are spherically shaped clusters of stars but Shapley realized that a high proportion of the clusters were host to a special type of star.

These special stars are now known as Cepheid Variables and they have a very particular property. Firstly, as 'variable' implies, the amount of light they give off varies, but more interestingly there is a correlation between how long they take to change their brightness and how bright they are at maximum output. Knowing this, if I were to measure how long the star takes to fade from brightest to dimmest, I would be able to infer how much light it is actually giving off. From this, I could measure how bright it seems in the sky and, because light fades over distance, I could work out how far away it is. Shapley knew this and calculated the distances to hundreds of Cepheid Variable stars and therefore the globular clusters themselves.

He found the clusters seemed to be roughly distributed in a spherical halo but this halo was not centred on our Sun; instead it seemed to be centred on a point around 32,600 light years away, in the direction of the constellation of Sagittarius. He also deduced that the galaxy was about 100,000 light years in diameter. A light year is a unit of distance used in astronomy and is equal to the distance light can travel in one year at a speed of around 300,000km per second, a pretty big number by anyone's reckoning, so Shapley's estimate meant that light reaching the Earth from the galactic centre took around 32,600 years to reach us. He was not far off with his estimate, which was an incredible feat: the currently accepted figure is between 26,000 and 28,000 light years.

The final step in understanding our place in the Universe was made by an American astronomer, Edwin Hubble (after whom the Hubble Space Telescope is named). He, like Shapley, had been employed by the Mount Wilson Observatory in 1919 as a junior astronomer. In his time there, he set about studying the strange spiral nebulae – interstellar dust clouds – which appeared fuzzy in the sky and were thought to be star-forming regions in the Milky Way.

Using the recently completed 2.5m reflecting telescope, Hubble took a succession of pictures of the spiral nebulae, including the so-called Andromeda Nebula. By comparing pictures taken over successive nights he was able to spot changes and on 4 October 1923 he identified another of the Cepheid Variable stars, but this time it was in the Andromeda Nebula. Using the same method as Shapley, he was able to determine the distance to the star and hence to the nebula, and to the surprise of his colleagues, who all thought it would be within our galaxy, he found it to be 900,000 light years away and far beyond the Milky Way. By discovering and measuring the distance to other galaxies, Hubble had increased the size of the known Universe significantly.

Following on from his discovery, it was realized that the type of Cepheid Variable Hubble was observing was a previously undiscovered brighter sort, and comparing it to fainter ones in our own galaxy meant the estimated distance to Andromeda doubled overnight to almost 2 million light years. We now know from even more accurate methods that the distance to the newly renamed Andromeda Galaxy is 2.3 million light years.

Hubble did not stop there though. He busied himself to measure the speed and direction of the motion of the galaxies, continuing the work of his fellow American astronomer Vesto Slipher just ten years earlier. He did this by splitting the incoming light from the galaxy into its component parts using a spectroscope, much like water drops split incoming sunlight into a rainbow. By measuring certain properties of the spectrum, Hubble deduced that some galaxies, such as Andromeda, were moving towards the Milky Way while others were moving away. This was true of the nearby galaxies, but at larger distances it seemed everything was moving away from us at quite incredible speeds. This either meant that the Milky Way was the centre of the Universe or that Hubble had discovered a general expansion of the Universe. A great way to visualize this is to imagine you live on the surface of a balloon and as it is blown up all points on the surface seem to rush away from you. There is no centre of expansion; it is just that every part of the balloon is moving away from every other part.

It seems then that our galaxy, the Milky Way, is one of millions if not billions of galaxies, all of which are like islands in space, made up of stars, clusters, nebulae and even other planets. Hubble's discovery marks the end of our journey for this chapter and notwithstanding a few more recent discoveries broadly explains how we got to understand the general layout and structure of the Universe we see today.

It is interesting to look back at the way this view has

evolved since our ancestors crawled out of the caves. It is easy to understand why early civilizations thought that we sat at the centre of the Universe, since everything in the night sky wheeled overhead. Yet with every discovery came a slight blow to the human ego as we slowly but surely were dislodged from the centre of everything. First the Earth was found not to be the centre of the Solar System, then the Solar System was found to be in a fairly remote part of the Milky Way Galaxy, and finally the Milky Way itself was found not to be at the centre of the Universe but instead just one of countless other galaxies all racing away from each other.

In many ways, the journey of understanding the Universe has also been a journey in understanding our place in it. As science progresses and new discoveries are made this view will undoubtedly change, but I am pretty sure the general overview presented in this chapter will stand the test of time. The last few thousand years have been a testament to human ingenuity and courage as great leaps were made often against much opposition, but as you continue in your journey around the night sky, remember you are not just looking out into space but also looking back through the history of mankind.

January: Northern Hemisphere Sky

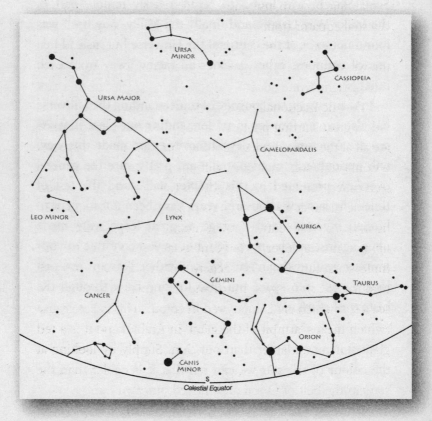

Looking at the sky it is easy to see why our ancestors believed all stars were fixed onto a giant crystalline sphere surrounding the Earth. The distinctive stars in the constellation Orion are a great example of what might have contributed to this illusion and it is here that we start the January sky guide.

Remember to read the section starting on page 16, explaining how to use these guides and how to find the celestial equator from your location as all guides start there. Once you are looking roughly in the right direction, just to the west of your gaze should be an almost horizontal line of three bright white stars. These are the stars in Orion's belt and they lie fractionally below the celestial equator and provide a great starting point. At first glance they look like they are all at the same distance from us, but in reality the stars of Orion's belt each lie at quite different distances. The eastern star, called Alnitak, is 817 light years away, which is a comparable distance to the western star, called Mintaka, at 916 light years. The central star, Alnilam, is a monstrous 1342 light years away, over 50 per cent as far again as Alnitak. The three stars can easily be seen to be shining with a white colour. Move your gaze north and fractionally to the east to see a star which is distinctly red in colour. This is Betelgeuse (which means 'armpit of the giant' in Arabic) and it is a red giant star much larger than our Sun. Simply by looking at the colour of the star we can tell that it is cooler than the stars in the belt of Orion and indeed our own Sun.

Following a line from Betelgeuse due north-east takes you to the constellation of Gemini with its two most prominent stars, Castor and Pollux, marking its north-east border. Castor is the white-coloured star to the north with Pollux to its south looking a little more yellow. Roughly halfway between Pollux and Betelgeuse is Gamma Geminorum, which is the third-brightest star in the constellation. Taking a line from Gamma Geminorum to the north-west takes you to Elnath in Auriga. Scan the area of sky halfway between the two with binoculars and you will notice not only lots of stars from the Milky Way but also an area where the stars seem particularly dense. This is M35, a not especially catchy name, but many galaxies, clusters and nebulae do not have real names but catalogue numbers instead, M35s coming from the Messier catalogue devised by the astronomer Charles Messier in the eighteenth century. Messier was actually a comet hunter and kept stumbling upon objects which were fuzzy and initially looked like distant comets, but they did not move so clearly were not. He catalogued them all so as not to waste time on them again and in so doing produced one of the most popular deep-sky catalogues in use today. Other catalogues exist, such as the New General Catalogue (NGC) and the Index Catalogue (IC), and many objects will have entries in more than one; for example, M31 is also known as NGC224. M35 is a stunning open cluster of stars with around a hundred stars visible through binoculars or small telescopes. Open clusters like this tend to be found inside our galaxy, unlike the globular variety used by Shapley to determine the shape of the Milky Way, which are found in a halo around the outside.

Almost directly due north from Elnath is a bright yellow star called Capella. In appearance it is much like our Sun, although its colour is the only similarity. Capella is the brightest star in the constellation of Auriga and to the naked eye appears to be a single star. In reality it is two binary star systems, four stars in all, just over 42 light years away. Within the boundaries of Auriga, which looks much like a misshapen square, are three other open clusters all found generally about 5 degrees to the north-east of Elnath. They are named M36, M37 and M38 and sweeping the area with binoculars will reveal them as tiny collections of glittering stars.

To the east of Auriga is the faint constellation of Lynx with Ursa Major and the famous collection of stars known as the Plough just beyond. Just under 10 degrees to the north of Castor in Gemini but still in the constellation of Lynx is the globular cluster known as NGC2419. It was discovered in 1788 by William Herschel and at magnitude 9.1 requires a telescope with an aperture of about 10cm to be seen. At a distance of 300,000 light years it is one of the most remote globular clusters of our galaxy.

Starting from Gemini again and moving south is another example of a star which, like Capella, is in fact a binary system. It is found by looking to the east of Betelgeuse, at a bright star which looks yellow-white in colour. This is Procyon in the constellation of Canis Minor and it lies 11.4 light years away. The larger of the two stars, Procyon A, is twice the diameter of the Sun and the smaller, Procyon B, is a white dwarf star one tenth of the Sun's diameter.

Starting from the belt of Orion again, look over your right

shoulder and you will see the Great Square of Pegasus setting in the west. It is part of the constellation of Pegasus,
representing the winged horse, and at this time of year it is
up on its corner looking more like a diamond than a square.
The northernmost star, called Alpheratz, is the one we are particularly interested in as we can use it to 'star-hop' to a real
treasure and one that will take you into deep space. From
Alpheratz, which is a star also shared with the constellation
of Andromeda, continue moving north, slightly to the east and
away from the horizon into the constellation of Andromeda.
The line bends slightly past the first star (Delta Andromedae)
and stops at the second, which is the second-brightest in the
constellation. Now you have to be very careful, as you need to
find two quite faint stars, which you can discover by turning
westwards as though you were heading towards the north-
west horizon. You will find one star which is fainter than the
one you have just left and another even fainter one that is only
just visible to the naked eye in a good dark sky. To the west of
this star is a faint fuzzy patch. Binoculars will show it to be a
slightly bigger fuzzy patch, and not surprisingly a telescope
will reveal it to be a much bigger fuzzy patch. What you will
also see though, if you have a large telescope and the skies are
dark, are dark lanes against the smudge of light. It may not
look incredibly spectacular but, as photographs show, it is a
vast swirling mass of stars called the Andromeda Galaxy, the
nearest major galaxy to our own at a staggering distance of 2.3
million light years. In other words, because of the time it
takes light from the Andromeda Galaxy to get to Earth, we
are seeing it as it was over 2 million years ago.

January: Southern Hemisphere Sky

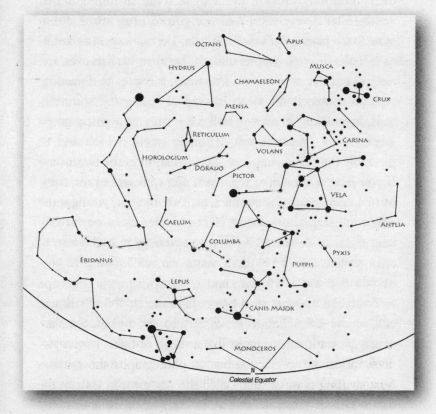

Glance along the celestial equator in the January southern sky and you will see the stars in the constellation of Orion are prominent. The three belt stars, Alnitak on the eastern side, Alnilam in the middle and Mintaka on the western side, are prominent white stars. We start this guide from the central star, Alnilam, and imagine a line dropping to the south. After a very short distance you can just make out a faint fuzzy patch, the Orion Nebula. Do not expect to see it as it looks in photographs though, because human eyes are not as sensitive as a camera lens when it comes to detecting colours if objects are faint. Instead of seeing the stunning reds, greens and blues you will see it only as a wispy grey-green object, but it is still definitely worth a look and is probably the best example of star formation in the sky today. If you have a telescope, see if you can spot any of the stars buried deep inside the nebula which are the hot, young stars called the Trapezium cluster. Infra-red telescopes are used to peer deep inside the dust cloud to reveal its hidden secrets. At a distance of 1344 light years we are looking at the nebula as it was 1344 years ago.

To the south-west of the nebula is the bright star Rigel, still in the constellation of Orion. To the naked eye this bright blue supergiant looks like a single star, but telescopic observation reveals a companion star 500 times fainter. Further studies of the light of Rigel's companion star reveal that it is a very special sort of binary star called a spectro-scopic binary. This means Rigel's companion is visible only when the spectrum of its light is studied and close examin-ation reveals superimposed lines that are the tell-tale sign of particular gases. The movement of the two stars is shown by the movement of the lines against the background spectrum, first towards the blue end and then towards the red as the companion moves first towards and then away from us.

Another bright star is easily spotted close to Rigel, further over to the east, and it is called Sirius in the constellation of Canis Major. Sirius is one of the brightest stars in the sky, rivalled only by the Sun, Moon and planet Venus. It is another binary star system, consisting of two white stars, the brighter one Sirius A and the older and fainter Sirius B. The companion star, Sirius B, is faint at magnitude 8.5 and due to the proximity and brightness of Sirius A is incredibly difficult to observe. Heading a little further south another bright white star is visible, Canopus in the constellation of Carina. This constellation is famous for a nebula that surrounds a star further over in the east, Eta Carinae, which is a binary star with a Wolf-Rayet star, which is one that is losing mass rapidly as a result of stellar winds, in orbit around a larger companion. Just to the east of Canopus is another bright and unmistakable star of the southern hemisphere, Gamma Velorum, or Suhail al Muhlif, meaning 'Glorious Star of the Oath' in Arabic. The name Gamma Velorum suggests it is the third-brightest star (gamma is the third letter in the Greek alphabet) in the constellation of Vela, although in reality it is its brightest star. It is a complex multiple star system, the brightest member of which is a spectroscopic binary.

A little further south lies the South Celestial Pole but it is not as easy to locate as the northern celestial equivalent. To find it, look for a group of stars called Crux, or the Southern Cross, which lies to the south-east of Canopus. From there it is easy to follow the stars forming the longer axis of the cross to the South Celestial Pole, which unfortunately is not marked by a bright star, unlike its northern counterpart.

The Southern Cross actually lies along the same line of sight as the plane of our galaxy, the Milky Way, which can be seen arching across the sky from the south-east to the north-west. The band of light we can see making up the Milky Way arises from the combined light of an estimated 400 billion stars.

Looking at the Milky Way with the naked eye, particularly from a dark location, reveals many dark lanes, which are intergalactic dust clouds, but you can also see brighter regions. One particular bright patch of fuzzy light is found to the south of the bright star Canopus, but it seems a little disconnected from the main band of light. This is the Large Magellanic Cloud. This is not an extension of the Milky Way but, instead, is one of the two main satellite galaxies to our own, at a distance of around 160,000 light years. It straddles two constellations, the long and slender Dorado to the north and the much fainter constellation of Mensa to the south.

In 1987 a star in the Large Magellanic Cloud, just to the edge of the Tarantula Nebula, exploded as a supernova, one of the most violent events in the Universe. Before the event was observed visually, a burst of neutrinos (particles with no electrical charge and difficult to detect) was recorded, which it is thought was emitted at the precise moment of core collapse. Imagine a line from the Large Magellanic Cloud to the north all the way to Rigel, and directly between the two is a globular cluster which lies in the faint constellation called Columba. The cluster has the catalogue designation NGC1851 and at 7th magnitude is just about visible with binoculars from a dark site. It lies 39,000 light years from our Solar System and around 54,500 light years from the centre of the Milky Way.

Forming a great triangle in the sky with Rigel near the celestial equator and Canopus to the south is the brightest star in the constellation of Eridanus, called Alpha Eridani. Also known as Achernar, it is a massive star containing about eight times as much material as our Sun and it is about seven times as big. It marks the southernmost tip of the constellation, which is depicted as a great river stretching from the most northernly point of Hydra, weaving through the sky to Cursa, the northernmost star, which lies close to Rigel in Orion.

TWO

Navigating the Night Sky

IT IS VERY EASY for me to rattle off a whole host of fascinating objects to look at in the sky, point out their location and even steer a high-powered telescope to reveal them in their full glory. Modern amateur computerized telescopes can even point to objects for you but, for the beginner just starting out, equipment like this may be a little daunting, and certainly a little pricey. Other technological aids are available such as smartphones and tablet computer devices which run applications that can guide you around the sky. These tools are all impressive but there will be times when you need to find things yourself. If you are out and about with friends it is great to be able to find your way around the sky and show others little celestial treats.

Finding your way around the sky is not just a neat party trick though, as a thorough understanding of how the sky moves and knowing roughly where things are will prove invaluable to your observing sessions. The great thing is that while it may seem daunting, navigating the stars is not actually that difficult. There are some good techniques and tools that will help you in tracking down that faint galaxy or planet, and with a bit of practice you'll be whizzing around the sky like a pro. Before launching into these tips we first need to take a look at how the stars and objects in the sky move and at the coordinate systems used in astronomy.

We all recognize that things in the sky rise in the east and set in the west; from the Sun to the Moon and distant, faint galaxies, they all follow this daily ritual. The general movement of objects across the sky is caused by the rotation of the Earth as it spins on its axis once every day. This is why many advanced astronomical telescopes have drives attached to them that turn the telescope in the opposite direction to the spin of the Earth, which effectively stops the motion of the object being looked at. The rotation of the Earth coupled with the movement of the Earth around the Sun mean we see a different set of stars at different times of the year.

Exactly which stars you can see in the sky is determined not only by the time of year but also by your position on the Earth; for example, if you lived at the North Pole you would be looking out into space in one direction, but if you lived at the South Pole you would be looking in the opposite direction and see different stars. Because the Earth is almost a sphere and rotates on an axis, this also means that someone at the North Pole will never be able to see any of the stars that someone at the South Pole can see. This changes at the equator though, as there it is possible to see all the stars in both northern and southern parts of the sky over the period of one year.

If you observe the sky from a latitude between the poles and equator you will see some stars from the northern sky and some from the southern sky. If you consider again someone living at the North Pole, all stars would be constantly above the horizon, never rising and never setting, and it is the same at the South Pole. At the equator things are different, with all stars rising and setting and none staying above the horizon over a 24-hour period. At latitudes in between, some stars rise and set while others stay constantly above the horizon; the stars that stay above the horizon twenty-four hours a day are called circumpolar and, while their positions change, they will always be visible.

In order to be able to navigate around the sky or to be able to communicate the position of an object we need some sort of coordinate system for the sky, and two of these are common in astronomy. The first uses an object's altitude and azimuth to explain its position in terms of a measurement called a degree. The altitude part of the coordinates explains how high the object is above the horizon when the horizon is 0 degrees altitude and the directly overhead position (or zenith) is 90 degrees, and halfway between the horizon and overhead is 45 degrees. It is easy to approximate angular measurements like this on the sky and a great technique using your hand is explained on pages 17–18.

The other part of the coordinate system, azimuth, explains how far around the horizon you need to look, starting from the north. If you do not have a compass to identify north you can note which direction the Sun sets and when you stand facing that when it is dark, your right shoulder will be pointing north and your left shoulder facing south; this is regardless of which hemisphere you happen to be observing from. Alternatively, if it is dark and clear and you are in the northern hemisphere the star Polaris lies due north, or if in the southern hemisphere you can use the constellation called Crux (the Southern Cross) and a couple of bright stars in Centaurus to find the south and from that the north. From the north, wherever you are, azimuth defines a number of degrees around the horizon in the direction moving through east to your chosen target. You can imagine a great circle lying around the horizon, 360 degrees all the way round, with north at 0 degrees, east at 90 degrees, south at 180 degrees and west at 270 degrees. Using altitude and azimuth it is easy for positions of objects to be communicated. The only problem is that objects move across the sky as the night progresses so their altitude and azimuth will also change.

This problem is solved with a slightly more complex coordinate system whose components are called right ascension and declination. Let us consider the Earth, which is a sphere that has a grid coordinate system superimposed on it whose components are called latitude and longitude. Latitude defines position on the Earth north or south of the equator: the equator is 0 degrees latitude, the North Pole is 90 degrees, the South Pole is −90 degrees, and locations between the equator and the poles, like the UK, are at latitudes between 0 and 90 or −90 degrees.

The celestial equivalent to latitude is called declination and to understand how it works, imagine the Earth is surrounded by a great celestial sphere upon which everything in the sky is fixed. Extending the Earth's equator out onto the celestial sphere gives an imaginary line known as the celestial equator which reaches around the sky. If you stood at the Earth's equator the celestial equator would pass directly over your head, or if you stood at the South Pole it would lie around the horizon.

Like our own terrestrial equator and latitude, the celestial equator is the starting point for the celestial coordinate referred to as declination and defines position in the sky north or south of it. The celestial equator itself has a declination of 0 degrees, the North Celestial Pole is 90 degrees and the South Celestial Pole −90 degrees. The star called Polaris is very close to the North Celestial Pole and has a declination of just over 89.5 degrees. As you can see from the declination of Polaris, it is sometimes necessary to measure angles smaller than a degree. Each degree can equally be divided into 60 segments called minutes of arc (arc-minutes), which can themselves be divided into 60 equal seconds of arc (arc-seconds). While it is perfectly fine to say that Polaris is at 89.5 degrees declination it is also accurate to say 89 degrees and 30 arc-minutes (30 arc-minutes being equal to half a degree).

Back to the Earth and our terrestrial coordinates: if latitude defines position north/south, the other half of the system is longitude, which defines position east/west of the Greenwich Meridian – also known as the Prime

Meridian – that runs through the UK, France, Spain and five other countries. Heading east, longitude increases up to 180 degrees, which is on the opposite side of the Earth, and moving west from the Meridian, longitude decreases to −180 degrees, although practically speaking these lines of longitude are the same.

The starting point for longitude was essentially the result of an arbitrary decision on political grounds but the starting point for its celestial equivalent, right ascension, is a little more scientific. We've already looked at the celestial equator and its origins but there is another imaginary line on the sky. The Earth moves around the Sun, taking a year to complete one orbit, but because we live on the Earth we perceive this as the Sun moving against the background constellations, so for example, in February it lies in Aquarius but in March it will be in Pisces. The line that the Sun traces against the sky is called the ecliptic and it is angled to the celestial equator by just over 25 degrees. There are two points where these two imaginary lines cross and one of them is the starting point for right ascension. It is called the First Point of Aries, although due to the wobble of the Earth on its axis it actually lies in the constellation of Pisces.

Unlike longitude on Earth, which is measured in degrees, right ascension is measured in hours going east from the First Point of Aries all around the sky, and twenty-four hours brings you full circle back to the starting point. It may seem confusing to use hours instead of degrees but imagine standing at the equator when the First Point of Aries is directly overhead. An hour later it will have moved towards the east by an angle equal to 15 degrees, so one hour of right ascension is equal to 15 degrees. An hour later it will have moved another 15 degrees, and so on, until just under twenty-four hours have passed, when it will have returned overhead again. Just as degrees are broken down into arc-minutes and arc-seconds, an hour of right ascension can also be broken down into smaller sections. As you might expect, one hour of right ascension is divided into 60 equal minutes and each minute into 60 seconds. Do remember, though, that 1 minute of arc is different to 1 minute of right ascension.

This is not all just theoretical stuff: the two systems, altitude/azimuth and right ascension/declination, both have important and practical uses in astronomy. The appearance of satellites or the International Space Station, for example, will be described in terms of altitude and azimuth, as will planetary positions most of the time. For other objects, the positions are generally referred to in terms of their right ascension and declination. Having an understanding of the origin and nature of the coordinate systems is important as it makes using them much easier. Right ascension and declination are probably the hardest to grasp but when it comes to telescopic astronomy they are the most useful since detailed star charts use them.

The simplest form of star chart is a device called a planisphere, which is usually made from plastic. It takes the form of two discs that are joined at the centre and rotate against each other. On the bottom disc is a map of the sky and, around the edge, the days of the year and marks that show right ascension. You will also see corresponding lines of right ascension radiating out from the centre of the planisphere. Cutting across the lines of right ascension and running around the centre of the map are the lines of declination, and you will notice a dotted line marking the ecliptic, the line along which the planets, Sun and Moon will all generally be found. On the top disc will be a transparent window which represents the portion of the sky you can see and, around the edge, the time of day. Using it is simply a matter of lining up the current time with the current date. It is important to note that you must make sure you get a planisphere for your location as they are calibrated for different locations on Earth.

Planispheres are great for naked-eye astronomy and finding which constellations are visible, but they can also be used to find the approximate position of the planets. A table is usually printed on the back of the planisphere and shows a value in degrees for each of the naked-eye planets for different dates. Make a note of the number for the planet and date you are interested in, turn the planisphere over and look around its edge next to the right ascension scale and you will see degrees marked. Find the point where the number of degrees is the same as the number you took off the table and, from there, draw a line to the centre of the planisphere. Where the line cuts across the ecliptic, the path followed by the Sun and planets, is where the planet will be found.

A modern alternative to the planisphere is the smart-phone or tablet PC. These devices are fantastic little bits of kit as most of them can tell where you are on Earth and even detect which way is north, south, east or west. These features have been put to good use for astronomy, with a huge range of applications that can be downloaded and greatly aid finding your way around the sky. Because they know where you are and the date and time, if held to the sky they can show you what can be seen in any direction. Coupled with a nice graphical representation of the sky they are fantastic for the naked eye, brilliant for the beginner, and will have you picking out planets, brighter stars and deep-sky wonders with ease.

More detailed than a planisphere or a smartphone application is a proper star chart, which can be bought in either book format, loose leaf or sometimes even laminated. They show many more stars than a planisphere, but without the date or time reference, and are just detailed maps of the sky. They do still have the lines of right ascension and declination so you can read off an object's coordinates with considerable accuracy. Both star charts and planispheres represent stars so that you can tell how bright they are, with the fainter stars shown as smaller dots and brighter ones as bigger dots. On a planisphere, the brighter stars are usually joined up to show the pattern of the constellations, but this is not always done on a star chart. You will also find galaxies, star clusters and nebulae marked clearly. Star charts come in a range of printed colours but it is best to steer away from all but white on black or, preferably, black on white. Red torches are the best way to read a star chart at night because your dark adaption will not be affected, but most colours can be quite hard to read under red light, particularly red print, which will just disappear.

One skill that it is really useful to master and will help you move around the sky to find your chosen object is star-hopping. It is simple in principle and only takes a little practice before you can use it with ease, but the great thing is that it works with naked-eye, binocular or telescopic observations. The technique relies on identifying a bright star or object that it is easy to spot and is near the object you are trying to pin down. If you are observing with the naked eye you simply 'hop' your way around the sky from this bright star, moving from one star to the next until you get to the object you are looking for. A simple example of star-hopping in the northern hemisphere is finding the Pole Star, or Polaris, which is not really all that obvious, by first picking out the easily recognizable stars at the end of the pan of the Plough (part of the constellation of Ursa Major, it is often described as being like a saucepan) and then simply following the line from them north until you come across the next star, which is Polaris. Another good example is the route to the Andromeda Galaxy by first picking out the bright star at the north-east corner of the Square of Pegasus and then hopping further east by two stars and north by three to locate the galaxy just to the north-west. It is all about finding a known starting point and imagining lines between stars on a simple star chart to devise routes to your target. In many ways it is just like driving a car to a destination: to make your way there you note other places along the way and 'hop' from one to the other.

Star-hopping with binoculars or a telescope is a little

more tricky because you can see more stars now, with the added complication through telescopes that the image is sometimes upside down or back to front, meaning that 'up' in the field of view is actually 'down' in the sky or your 'left' is actually your 'right'. Sounds confusing? But just try it and you will be surprised how easy it is. A great tip to make telescopic star-hopping easier is to get a piece of clear plastic and draw a circle on it which represents the amount of sky you can see through the binoculars or telescope – its field of view. With binoculars the field of view remains unchanged, but with a telescope if you change the eyepiece the field of view will also change so you may want a few different circles for each of your eyepieces. If you are not sure how big your field of view is you should point the binoculars or telescope at an area of sky where there are lots of bright stars so you can work out how much is visible. You can then draw the right-sized circle to represent the field of view. Not only will you need to do this for each eyepiece but also for each different star chart that you are using since the scales will differ from one chart to another.

To use it, place the plastic circle over a bright star on your chart that you can easily find and is close to your target object. This will be your starting point, but first work out if the image you can see is upside down or back to front. You can even do this in daylight by just looking through your telescope at something in the distance and noting how it appears. Now look on the chart and find another moderately bright star in the direction of the object you are after, move the disc over it and then find it in the telescope.

Keep doing this, hopping from one star to another using patterns of stars until you get to your target. Sometimes there may be no stars in that direction so you may have to make big jumps, estimating how far to move from your chart and plastic disc.

Another method of finding things with a telescope relies on attaching something called setting circles to the telescope. Using these, it is important first to make sure the telescope is polar aligned (this process is covered in the equipment chapter). Once this is done, you need to point the telescope at a known bright star and centre it in the eyepiece. Then find the right ascension (RA) and declination (Dec.) of the star and turn the setting circles to read the coordinates of the star. To find the target object you need to look up its own RA and Dec. and then turn the telescope until the setting circles read the correct values. Hopefully, if it has been set up accurately, you will now have a telescope pointing at the object you wish to see.

As you will see in Chapter 3, the ultimate luxury is to have a telescope that knows where everything is and can even point at it for you. Sounds like the stuff of science fiction, but telescopes like this are readily available today. You will have to pay a little more money for them and they are more complicated to set up, but once ready for use, which generally involves telling the telescope where you are on Earth, setting the date and time, and calibrating it on a couple of stars, the motors will turn it to point to any object in its database that you choose.

Impressive gadgets aside, there really is no substitute for

understanding the sky and how objects in it move and learning your way around it. To home in on a neighbouring planet or distant galaxy yourself is a great skill to have and one that will stay with you for many many years.

February: Northern Hemisphere Sky

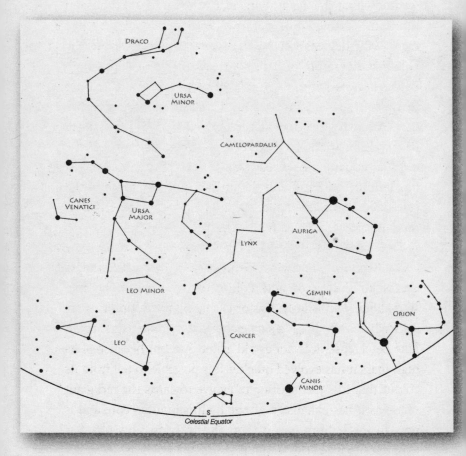

Looking along the celestial equator over to the south-west reveals the familiar constellation of Orion, which was introduced in January's guide. To the north-east corner of this famous constellation lies Betelgeuse, a prominent red giant star nearing the end of its life. Through studies of its spectra we know that this giant dying star is just under a thousand times the diameter of our own Sun and cooler at just over 3000 degrees kelvin, the scale used to measure the colour temperature of light sources. Surprisingly, although Betelgeuse is dying, it is only about 6 million years old (compared to the Sun, at about 5 billion years) but because of its original large mass it has evolved rapidly. It is said to have a declination of +7 degrees, putting it just north of the celestial equator. (As already discussed, declination is just like latitude on Earth and the further north or south of the celestial equator, the higher the number, with southerly declinations preceded by a minus sign.)

Due east and a little to the north of Betelgeuse lies the constellation of the twins, Gemini, with its brightest stars Castor to the north and Pollux to the south. The two stars differ significantly: Castor is a complex grouping of six stars whereas Pollux is a single cool yellow-orange giant star. Like Betelgeuse, it is not as old as the Sun but because of its high mass it has evolved quickly. It is possible to tell from its colour that it is just starting to evolve towards the red giant phase as it has exhausted all the hydrogen in its core and is now fusing helium into carbon and oxygen. This phase is expected to last 100 million years before intense stellar winds will puff the outer layers into space, creating a beautiful planetary nebula. Interestingly, it has at least one planetary companion, a Jupiter-sized planet called Pollux B, which will probably suffer a fate not too dissimilar to the Earth's when our Sun dies.

Just to the south-west of the twins Castor and Pollux in Gemini is a great example of what will happen to our Sun after it turns into a red giant. The object is too dim to be seen with the naked eye – in fact, there are no examples that can be seen without optical aid – but readers with a telescope should search out this instance of a planetary nebula, called the Eskimo Nebula. In photographs it looks just like the face of an eskimo surrounded by a fluffy hood and it is just about visible with a pair of binoculars. Like all planetary nebula, the outer layers of the original star were finally pushed off into space through the force of nuclear processes in its core, leaving behind a white dwarf star.

A little further along the celestial equator and just to the north of it is an easily spotted constellation which actually does look like its name suggests, Leo, the lion. To find it, look for a backwards question mark, which represents Leo's head, and extending out to the east is its body, including the bright star Regulus at the front and Denebola marking its tail. Just below the curve of stars marking its head lies the point in the sky where the Swift satellite detected a short ten-second burst of gamma radiation back in April 2009. This has the catalogue number GRB090423 and, as such, is one of the rather strange and not well-understood gamma ray bursters (GRBs). Fortunately for us, this one was at a distance of around 13 billion light years, so not close enough to cause us any problems. The burster is thought to have been generated at the moment a supermassive, fast-rotating star collapsed into a black hole at the far reaches of the Universe.

Directly to the north of Leo is the well-known group of stars called the Plough, or Big Dipper. This group is not an official constellation but instead part of a larger group of stars called Ursa Major, the Great Bear. At this time of year, the pan of the Plough, which represents the hind end and tail of the bear, sits high in the sky and horizontally around midnight with the rest of the constellation stretching out to the west and south.

At the western tip of the pan is another orange-coloured star, called Dubhe. Similar in nature to Pollux, it is a star nearing the end of its life and is rapidly evolving into a red giant. With the star at the southern tip of the pan, Merak, the two act as pointers to the North Pole Star, Polaris. Most of the stars that are seen in the Plough, excluding Dubhe, are part of a group of stars called the Ursa Major Moving Group, which consists of thirteen stars in Ursa Major and one in the neighbouring constellation of Canes Venatici to the south-east. The group is defined by stars that all share the same direction of movement as they hurtle through space, with the centre of the group about 80 light years away from us.

The second star in the Plough's pan handle from the eastern end is called Mizar and good eyesight will allow its fainter companion, Alcor, also to be seen. The two stars, separated by about one light year, are themselves binary stars, making this a quadruple star system. At the end of the handle which represents Ursa Major's tail is the star called Alkaid. Just to the north of this star is the Pinwheel Galaxy, which is a fine example of a spiral galaxy, whose orientation faces on to us, meaning its spiral structure is nicely visible through amateur telescopes. It also has the catalogue number M101, making it the 101st object in the catalogue published by Charles Messier in 1781 (see Chapter 1), and, like all galaxies, it is possible to split M101's light into a spectrum and determine how fast it is moving away from us. In the case of the Pinwheel Galaxy, it is moving away at 241km per second. The movement is not the result of the galaxy rushing through space but, instead, the expansion of space carrying the galaxy along with it. From studies of stars that have exploded as supernovae in distant galaxies, such as the one in the Pinwheel in August 2011, it seems that the rate at which space is expanding is getting faster rather than slower as previously expected.

Our knowledge of whether the Universe will continue to expand for ever or ultimately stop is determined by the study of galaxies and in particular of galaxies and galaxy clusters that act as gravitational lenses, such as HST14113+ 5211 just to the east of the Pinwheel Galaxy. It was discovered by the Hubble Space Telescope and is just one of many lenses that are caused by the presence of a galaxy cluster lying between us and a more distant one, a phenomenon that we shall explore in detail in Chapter 12.

February: Southern Hemisphere Sky

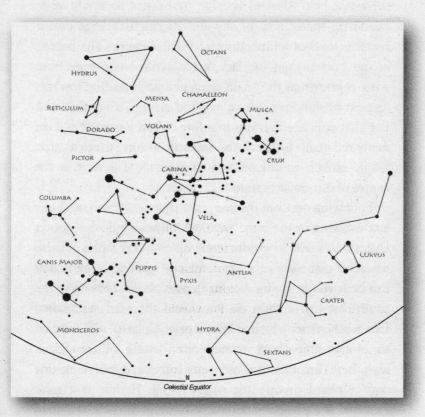

Hydra, the largest constellation in the sky, is well placed for observation in February yet it does not have that many bright stars. It starts just north of the celestial equator and stretches for quite a distance to the south-east. The brightest star in the constellation, Alphard, is found to the east of Sirius, which lies south of Orion and its famous three-star belt. Alphard is quite prominent as it sits in an otherwise sparse area of sky and its bright orange colour hints at its cool temperature of 4000 degrees. The orange colour comes from the fact that it is evolving away from what is known as the 'main sequence' and heading towards its next evolutionary phase as a red giant. This is the same fate that will face the Sun in a few billion years' time. Like many red giants it dwarfs the Sun and, at forty times its size, would stretch almost halfway to Mercury if it were at the centre of our Solar System.

Following on from the red giant phase, the Sun, and any star evolving in the same way, will ultimately lose its outer layers into space. There are many different examples of these so-called planetary nebulae in the sky but one in particular can be found in Hydra, not far from Alphard. Moving to the south-east just a short distance, find the next star, called Lambda Hydrae, which is fainter than Alphard, and then on to Mu Hydrae (both named after letters of the Greek alphabet). Together, the two stars form a slightly bent line with Alphard. Just to the south of Mu Hydrae is a great example of a planetary nebula; although not quite visible to the naked eye it is still worth hunting down with a pair of binoculars or a small telescope. Binoculars will reveal a star that looks slightly fuzzy, while a small telescope will show a blue-green disc whose apparent size in the sky is broadly similar to that of Jupiter, hence its name, the Ghost of Jupiter Nebula. Much larger telescopes are required to see the faint stellar core known as a white dwarf.

Just to the south of Alphard is a small, faint constellation called Antlia, which has no prominent stars in it, but a little further on, towards the South Celestial Pole, is another constellation, Vela, that was referred to in the January guide along with the prominent multiple star system Gamma Velorum. A little further to the west is the rather strangely shaped constellation of Puppis, which was once part of a much larger constellation called Argo Navis that represented a ship. It was divided in the 1700s into four separate constellations, of which Puppis is one. It is easy to locate as the Milky Way runs through the northern part of the constellation.

Within the borders of Puppis are a number of worthy deep-sky treasures to hunt out such as M47 (the 47th object in Charles Messier's catalogue). This stellar cluster is near the limit of visibility to the naked eye, so a dark observation site is needed. It is right on the northern border of Puppis and can be found by looking to the north-east of the bright white star Sirius in neighbouring Canis Major on the very edge of the Milky Way. To the south of Sirius is the bright yellow star Canopus and about halfway between the two, slightly to the east, is the orange star Pi Puppis, which is the rough position in the sky where meteors from the Pi Puppids meteor shower seem to come from. In reality the pieces of rock and dust are travelling through space in parallel streams, but the effect of perspective makes them seem to come from one point in the sky. The shower peaks in April but only when the comet 26P/Grigg-Skjellerup makes its closest approach to the Sun, every five years. Meteor showers generally dump hundreds of tonnes of meteoric dust onto the Earth every day but this is not enough to upset life. It is the larger lumps of rock that cause devastation, even extinction, as we shall see in Chapter 12.

Continuing to the east, the next star along is a little brighter and is the blue supergiant Naos, or Zeta Puppis. This star is a fairly rare breed and among the hottest stars in the Universe, with a surface temperature of around 42,000 degrees, compared to our Sun at just under 6000 degrees. To the north of Naos by half a degree (a finger held at arm's length represents a degree so half of that) marks the location of one of the closest GRBs ever detected, GRB031203.

Gamma ray bursters like GRB031203 are impossible to predict and are thought to appear at the moment when a supermassive star finally collapses into a black hole. Initially the burst is seen in gamma ray wavelengths but afterglows can be detected at longer wavelengths such as X-rays, visible light and radio waves. GRB031203 was picked up in December 2003 and studies showed that it happened at a distance of 1.3 billion light years from us, so the light has taken 1.3 billion years to reach us. GRB031203 was unexpectedly faint, which means it was either a different type of burster or was a different type of event, known as an

X-ray flash. These are not well understood and only a few have ever been observed.

Bordering Puppis to the south-east is the constellation of Carina with its unmistakable star Canopus as its brightest. Between Canopus and the Southern Cross high in the east lies the renowned star system of Eta Carinae, which is surrounded by the nebula NGC3372 (New General Catalogue). The nebula is visible to the naked eye and at a distance of around 8000 light years may pose a threat to the Earth. The nebula is thought to be home to one of the most massive stars in the Universe, which is orbited by a Wolf-Rayet star that is losing mass rapidly. The supermassive star Eta Carinae has gone through significant brightness variations since its discovery, which suggests instability leading ultimately to the star going supernova or maybe even hypernova, the most violent explosion in the Universe.

All of the objects in the sky guide this month have had roughly the same azimuth, i.e. the measurement around the horizon from due north, but they have differed in altitude as the guide has moved from the celestial equator to the South Celestial Pole. As the night progresses, the altitude and azimuth will change as the objects move from east to west.

THREE

Equipment

Whether you are new to stargazing and enjoying the night sky or a seasoned naked-eye observer, there will probably come a time when you want to take the next step and get a closer look at the celestial wonders you have been hunting down. To the inexperienced, the world of telescopes can be daunting, with a whole host of new terms and concepts that might be completely alien to you. Fear not, this chapter will teach you all you need to know, but perhaps it is worth considering first whether a telescope is the right tool for you as there is an alternative, cheaper option that will give you a closer look than the eye alone can offer.

Whether you should go straight for a telescope or not largely depends on your level of experience. I have seen newcomers spend thousands of pounds on a high-end instrument only to find astronomy is not quite for them: the result is wasted money. If you are a relative newcomer you may be better off buying a pair of binoculars first. If you then find stargazing is not for you, you have not spent lots of money and you can always use them for other more down-to-earth purposes.

All binoculars are described in a standard way, which is made up of two numbers, e.g. 7 × 50. This particular pair would offer a magnification of seven times (making objects appear seven times bigger) and have lenses that measure 50mm across. This last number is quite important as the lens diameter dictates how much light the binoculars will collect and therefore determines the faintest object you will be able to see. You can think of binoculars and telescopes as funnels for letting more light into your eyes than you would normally be able to detect.

When choosing binoculars it is better to have a bigger second number and therefore bigger lenses, and not so important to have a higher magnification as telescopes really are the instruments for higher-magnified views of the night sky. One downside to choosing binoculars with a larger lens or aperture is that they are larger and heavier and, while they will be magnifying the view of the night sky, they will also be responding to any movement your hands make. Tiny tremors or even your heartbeat will be picked up and magnified and will have an adverse effect on your view.

One solution to making sure you have a steady view is to mount the binoculars on a tripod. Many people own or have access to a camera tripod through friends or family and they can greatly enhance the experience of observing with binoculars. You will need to buy an adaptor that will fix the binoculars to the top of the tripod but these are not too expensive. The only downside is that fitting binoculars to a tripod makes looking at objects directly overhead very awkward if not impossible.

An alternative solution is to buy a pair of binoculars that use image stabilization technology. As its name suggests, this steadies the image you see through the binoculars. It will not mean you get a rock-steady image from a moving car but it will remove the slight hand tremors as you stand still and look through them. It is amazing how much more detail you can see when the image is nice and stable, so even if you do not want to spend the extra money on image stabilization, a tripod and adaptor are a worthwhile investment.

With a decent pair of binoculars it is possible to really open up the wonders of the night sky. They will not show a great deal of extra detail on the planets, although a higher-powered pair will just reveal a hint of the rings of Saturn and even the moons of Jupiter. The surface of the Moon will be enhanced nicely and you will be able to pick out smaller craters and finer detail in the lunar highlands, but it is not just the brighter objects that can be seen in greater detail. There are a whole host of deep-sky objects that now come well within your range. The March sky offers some great examples of easy-to-spot deep-sky objects for binocular-users. In the northern hemisphere sky the wonderful open Beehive cluster in Cancer is easy to find and is one of my favourites. Southern hemisphere observers can hunt down the Southern Pleiades, another stunning open cluster in the constellation of Carina. Both can just be detected with the unaided eye from a dark site but binoculars make them easy targets.

Eventually there may come a time when binoculars will not be enough and you will want to take an even closer look, so it is time to think about buying a telescope. If you have ever looked at telescopes on sale in either a department store or a specialist astronomy shop you will have seen a lot of phrases and words that probably did not mean all that much to you. Starting with the basics, there are three different types of telescope: refractors, reflectors and catadioptrics. Refracting telescopes have lenses inside and are of similar design to those you'll see being used by sailors in films; reflectors use mirrors and catadioptrics have a combination of the two.

Most telescopes, regardless of their design, have a number of components in common: a tube to hold everything together, a mechanism – either a lens or a mirror – for collecting incoming starlight, an eyepiece to magnify the image, a smaller 'finder' telescope and a mount to hold the whole lot steady. The exact detail of each of these components varies from one telescope to another and with some it is possible to mix and match, or you could even build your own.

The most common telescope for newcomers is the reflecting telescope (see illustration 1) as it offers more for your money, largely down to the optics inside. At the heart of a reflecting telescope is a circular mirror called the primary. It has a very accurate curve ground into its front surface, whereas inside a good-quality refractor (see illustration 2) are typically two or more lenses glued together, all of which have had curves ground into both sides. Much more work is needed to make the lenses compared to a single mirror and the glass has to be of higher optical quality. It is the greater production costs and quality of the refractors that drive their price up, so as a beginner consider reflectors as your first choice.

The primary mirror is the device that collects the incoming starlight in a reflecting telescope and the curve on the front surface of the mirror takes the incoming beams of light and focuses them to a point. The mirror sits at the bottom

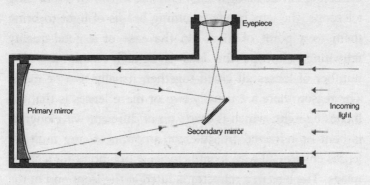

Illustration 1: Reflecting Telescope

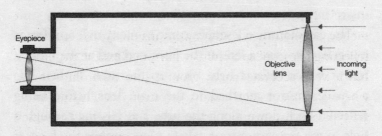

Illustration 2: Refracting Telescope

end of the telescope tube so it can reflect the incoming beams of light back up the tube. The converging beams are intercepted by another smaller, flat mirror called the secondary which is set at an angle at the top of the tube to direct the beams of light out of a hole in the side. It is at this hole that a special tube called a focuser is fitted to hold the eyepiece.

Refractors are designed differently, with the light-collecting device being a lens. Like the mirror in a reflecting telescope, the lens bends incoming beams of light to bring them to a point of focus. In the case of a good-quality refracting telescope the lens is actually made up of a number of lenses all glued together, usually two or more. The reason there are usually two or more lenses is that the incoming light, which is made up of different wavelengths, is bent or refracted by different amounts. Using multiple lenses ensures all wavelengths focus at one point for a sharp image. The lens in a refractor is fitted at the front end of the telescope tube so the starlight will pass through it and pass along the tube. On reaching the far end, the light passes through the focuser, where an eyepiece is held and the image formed.

The catadioptric telescopes come in many different forms but generally have a lens at the front end and at the other a mirror which reflects light back up the tube. It then hits a smaller mirror attached to the main lens before being reflected back down along the tube and passing through a hole in the primary mirror, where it encounters the focuser and eyepiece.

An important aspect of maintenance for telescopes, more so for reflectors and catadioptrics than refractors, is the alignment of the optics. If they are not aligned accurately then there will be a significant reduction in the sharpness of the image seen. The process of aligning telescope optics is called collimation; rough collimation can usually be done in daylight with more precise fine tuning achieved by pointing the telescope at a moderately bright star at night. Due to the way the lenses are held they generally do not lose their alignment, but mirrors are susceptible to movement, so reflecting and catadioptric telescopes should be regularly checked for collimation. This is particularly true if you often transport your telescope to different observation sites.

There are many different types of telescope on the market, most of which have pros and cons, but all of them are described using the same terms. The two most important are aperture and focal length. Aperture refers to the diameter of the main lens or mirror and gives an indication of the telescope's ability to collect light. The larger the aperture, the more light it can collect, and therefore the fainter the objects that can be seen.

A fairly standard beginner's telescope will have an aperture of around 150mm and will show objects around a million times fainter than can be seen by the naked eye alone. There is a practical implication to this: if you want to see very faint objects like galaxies and star clusters, you will need a large-aperture telescope, whereas if you are hoping to track down brighter objects, like the planets, a big aperture is not of quite so much importance. For example, if you want to take a look at the galaxies M81 and M82 in Ursa Major then they will certainly be visible with a small 150mm telescope, but telescopes larger than this will collect a lot more light and reveal much more of the galaxies' structure.

Not only does aperture determine the limit to which faint objects can be seen but, because it defines how much light a telescope collects, it will also define how much you can magnify the image. Magnification refers to how big an object appears compared to a naked-eye view, but as magnification gets higher so the image gets slightly darker. How high you can push the magnification depends on how much light there is in the first place, so a telescope that collects more light, i.e. has a larger aperture, will be able to take a higher magnification. To determine the maximum practical magnification a telescope can produce is a simple matter of multiplying the aperture in millimetres by two, so a 150mm telescope would give a maximum practical magnification of around 300x – any more than this and the image would generally become too dark.

Once the light has reflected off the mirror or been refracted through the lens it travels into another optical device, the eyepiece, which is a small tube that slots into the focuser of the telescope. Inside it is a series of lenses whose purpose is to magnify the image by an amount that is determined by something called focal length. The focal length is the distance it takes for incoming beams of light to be focused to a point and, perhaps obviously, a shorter distance means a shorter focal length.

Given that magnification is determined by focal length,

changing the eyepiece will actually give you a higher or lower magnification, but the exact amount achieved is calculated by dividing the focal length of the telescope by the focal length of the eyepiece. For example, if a telescope has a focal length of 1000mm and we are using an eyepiece that has a focal length of 20mm, the magnification would be 1000 divided by 20, which is 50x. You will see that an eyepiece with a longer focal length will produce a lower magnification than an eyepiece with a shorter focal length. The easiest way to remember this is that an eyepiece with a big number means a smaller image and an eyepiece with a small number means a bigger image.

It is easy to draw the conclusion from all this that as a new telescope owner you need to have a number of different eyepieces to give a range of different-sized images. The eyepiece you choose will depend on the object you are observing, e.g. low power for deep-space objects and high power for planets. You will also find that not only will the telescope determine which eyepiece you can use but the weather will often force you to use a lower magnification because of unsettled conditions. My eyepiece collection consists of eyepieces of various brands but has a good spread of focal lengths at 56mm (53x), 24mm (125x), 12mm (250x) and 6.4mm (469x), but of course on a different telescope you would get different magnifications. A good starter set might be a 24mm and a 12mm eyepiece or similar. You can also get devices called Barlow lenses which double, triple or even quadruple the magnification achieved with an eyepiece; for example, using a 24mm eyepiece on my telescope gives 125x magnification and I could use it with a 3x Barlow lens and get a magnification of 375x. These are a good addition to your kit as they will effectively double the number of eyepieces, but plan carefully as you will not want to duplicate your magnifications.

As with telescopes, there is a huge range of eyepieces on the market of different makes and focal lengths and prices vary significantly. If you are new to the subject it is best to go for medium-priced eyepieces as you will still be getting reasonable quality but not breaking the bank. Newcomers often make the mistake of buying a good-quality telescope followed by cheap eyepieces, but be warned: a cheap eyepiece can totally ruin an otherwise fantastic view.

One final word about eyepieces. There is a term called eye relief, which is the distance your eye should be from the eyepiece to get the optimum view. This is of particular importance when it comes to observing if you wear glasses. Short- or long-sighted observers can remove their glasses and adjust the focus of the telescope to suit. If you suffer from astigmatism then you will need to keep your glasses on so a long eye relief will allow you to observe more comfortably.

There is another telescope term called the focal ratio, which is written as, for example, f/x. It brings together the focal length and aperture of a telescope and its value is reached by dividing one by the other. Using the 150mm telescope earlier described, if it has a focal length of 900mm then its focal ratio is 900 divided by 150, which is 6, i.e. f/6.

Before looking at the mount that telescopes are fitted to it is worth just briefly mentioning the mini-telescope that fits to the side of a main telescope, called a finder telescope. Because the main telescope magnifies the sky by around 50x or more, it is incredibly hard to find your way around. For that reason a smaller, less powerful telescope is fitted to its side; it usually magnifies around 10x, giving a much wider field of view. By lining up the two telescopes so they are pointing in exactly the same direction, it is a simple matter to find an object with the finder telescope (or its location, as it is not always possible to see the object itself through the smaller finder) and it should be centred in the main telescope. This will take away the frustration of locating objects in the sky. Modern alternatives to the finder telescope are available which project a tiny red dot onto the sky and by aligning this to the main telescope you can again easily home in on your target.

We've now covered the optical components – telescope, eyepieces and finder telescopes – but none of these will be any good if they are fixed onto flimsy, wobbly mounts. A mount is the word used to describe the thing the telescope sits on and generally it is more elaborate than a camera tripod.

Telescope mounts come in two basic types, alt-azimuth and equatorial. A normal camera tripod can be considered to be an alt-azimuth mount as it moves around two axes: up and down (in altitude) and left and right (in azimuth). An astronomical mount of this type is not much different in that it moves around the same axes but in design and rigidity they do differ. A classic example of this type of mount is the

Dobsonian style, which is a rather strange-looking box device that the telescope sits in, designed by the American amateur astronomer John Dobson. Just like the camera tripod, it allows the telescope to move in altitude and azimuth and is probably the best telescope mount for a beginner as there is minimal set-up time and it is very easy to use.

One downside with using an alt-azimuth mount is that it is difficult, though not impossible, to attach motors to enable it to follow objects across the sky. When objects in the night sky move their position changes in both altitude and azimuth as they arc across the sky. To follow them, it is necessary to fit a motor to both axes of the telescope and this will allow you to freeze their motion. It will not, however, stop them rotating in the eyepiece. As they rise and set they follow a curved path and as a result appear to slowly spin while we look at them. This apparent rotation is not an issue in visual observation, but if you want to try your hand at photography it becomes a problem, as exposures can last for some minutes and in that time the object will have rotated, leaving you with a blurred image.

The solution to this is a different type of telescope mount, the equatorial mount, which differs from the alt-azimuth style in the orientation of the axes. With the alt-azimuth mount, a vertical axis allows the telescope to swing horizontally around the horizon, in azimuth. With the equatorial mount the same axis (the polar axis) is tilted so it is parallel to the Earth's axis of rotation. This means that as the telescope is moved around that axis, it follows the same arc across the sky as the objects. With this type of mount it is possible to attach a motor to just one axis, turn the telescope in the opposite direction to the Earth's rotation, but at the same speed, and objects remain in the centre of the eyepiece and do not spin. This means you need only one motor to follow or track objects across the sky and also allows for long-exposure astronomical photography without any blurring.

This all sounds fantastic, but the key in getting an equatorial telescope to accurately track objects across the sky is in a process called polar alignment. As its name suggests, polar alignment is the act of aligning the polar axis to the axis of rotation of the Earth. Advanced amateur telescopes will have small telescopes within the polar axis allowing it to be aligned to the north or south celestial pole. The mounts are designed so they will work for any location on Earth but they need adjusting for your specific location. Very rough polar alignment can be achieved by aligning the axis north–south and setting the adjustable angle of the mount to be the same as your latitude. If you are observing visually then that will be enough to keep the object in the field of view for a short while, but for astronomical photography more precise methods like 'drift alignment' are needed; that is too detailed for this book, but further information is readily available on the internet.

My choosing the right telescope system for you is a bit daunting, but for newcomers I would always recommended a 152mm or 200mm Dobsonian reflecting telescope around f/6 or f/7, which will be a pretty good all-round telescope and not cost a crazy amount of money. In your search for a telescope you might also see computerized telescopes that have the ability to control the motors and point at objects for you. These are fantastic telescopes but unless you are prepared to spend lots of money you will find the optical quality of the entry-level computerized instruments to be lacking. I would suggest steering away from them as your first telescope and upgrading to one in a few years.

One final piece of advice: before you spend your hard-earned cash pay a visit to your local astronomical club. It is one thing to understand the words written in this chapter but nothing quite beats the experience of seeing telescopes for real and taking a look through them. It is not just the price you should consider when making your purchase but portability, future expansion, potential upgrades and build quality, so seeing them for real and speaking to owners will soon have you homing in on the ideal telescope for you. There are plenty of specialist astronomical equipment suppliers in most countries, and they are also a great place to go for advice. It is best to keep away from the telescopes in department stores unless you are buying for children as they are pretty poor quality. Take your time though, try a few out with friendly local astronomers, and you will eventually know which one to buy and, who knows, before long you will be giving advice to other newcomers and sharing your experience.

March: Northern Hemisphere Sky

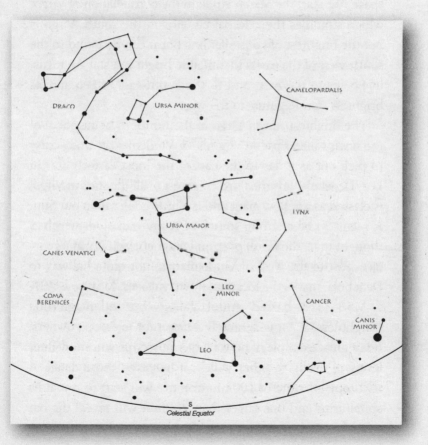

March is a great time of year for getting out under the stars and for picking out faint fuzzy objects like clusters and galaxies, because the obscuring stars and dust of the Milky Way run low around the northern part of the sky. This gives us an unimpeded view out in the other direction, into deep space. We start the March guide in the constellation of Virgo, which straddles the celestial equator in the south. Virgo is not the brightest of constellations but it can be found to the south-west of the easily identifiable bright red star Arcturus, in Boötes to the east, and to the south-east of Leo and its brightest star, Regulus, to the west.

The brightest star in Virgo in the northern hemisphere of the sky is called Epsilon Virginis, or Vindemiatrix, and is easy to pick out as it lies to the east of the most easterly star in Leo, Denebola. It is the third-brightest of all the stars in Virgo, is classed as a yellow giant and is a little cooler than our Sun. It seems to be a strong source of X-ray radiation, which is thought to be the result of strong magnetic activity at its surface. Just to the west of Vindemiatrix, not quite halfway to Denebola, marks the location of two galaxies, M58 and M87.

M58 is a barred spiral galaxy but at magnitude (brightness) 9.7 it is definitely a target for telescope owners. It is a great example of how a larger telescope will show finer levels of detail; in other words, it 'resolves' more detail. A telescope aperture of 100mm or more will start to reveal its spiral arms and one larger than 200mm will reveal the bar structure. It lies a staggering 68 million light years away from us and, of the 2000 or so members of the Virgo Cluster of galaxies, it is among the brightest.

Just to the west by about 5 degrees in the direction of Denebola is the largest, brightest and most dominant member of the Virgo Cluster, M87. This goliath of an elliptical galaxy is just visible with binoculars and easily seen through telescopes. Visually the galaxy does seem large and astronomical images show that it appears in our sky larger than the full moon, which is due partly to its proximity to us, at an estimated 52 million light years.

Moving directly north of Virgo is the fainter constellation called Coma Berenices and one of its brightest stars, Diadem, is found just to the north of Vindemiatrix in Virgo. There are only three main stars in Coma Berenices, arranged in a right-angled triangle, but there are many more fainter stars. The majority of the fainter stars make up part of a stellar cluster called Melotte 111, which is only 270 light years from us. Like Virgo, Coma Berenices is rich in galaxies; of particular note is the edge of spiral galaxy NGC4565 just south of Gamma Comae Berenices, the star on the constellation's western border. Telescope owners should search out this galaxy and, with a 200mm telescope or larger, its dark dust lane can be seen running along the disc of the galaxy.

To the north of Coma Berenices is another small constellation, called Canes Venatici, which has only two prominent stars in it. They can be easily found as they are parallel to the last two stars in the handle of the Plough to the north and point directly to Arcturus, the bright red star in Boötes. On a line between the brighter of the two Canes Venatici stars, Cor Caroli, and the last star of the Plough's handle, Alkaid, is the spectacular galactic collision known as the Whirlpool Galaxy, or M51. Starting from Alkaid, it is found about a quarter of the way to Cor Caroli and can just be seen in binoculars together with NGC5195, the galaxy it is interacting with. It is thought that NGC5195 passed through the main disc of M51 from behind around 500 million years ago before returning, for another 'collision' 400 million years later, to its present position, slightly behind M51. The two are believed to be entwined in a gravitationally bound dance which is likely to lead to an ultimate merger.

Just the other side of Alkaid, at roughly the same distance as M51, lies another fine example of a spiral galaxy, M101. Observations of Cepheid Variable stars inside M101 have allowed its distance to be calculated at 27 million light years. It is just beyond the limit of visibility to the naked eye so binoculars are needed to pick it up from dark skies, but a telescope of 100mm or bigger will start to show the structure of the spiral arms.

The stars of the pan of the Plough act as pointers to a lovely pair of galaxies. Imagine a line between Phecda at the south-east corner of the pan to Dubhe at the north-west corner and extend the line on to the north-west for the same distance. Scanning the sky in that area with binoculars will reveal the stunning contrast of the symmetrical spiral galaxy called M81 and, in the same field of view, the ragged form of M82, an irregular galaxy. There are many other fine examples of galaxies in the March sky, particularly around Virgo and Coma Berenices, so do not be restricted to the objects already covered – have a wander around the sky and discover more galactic surprises.

March: Southern Hemisphere Sky

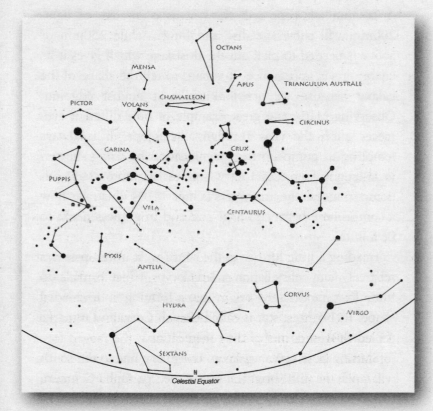

Spica, the brightest star in the constellation of Virgo, is dominant in the southern hemisphere sky during March and it provides a useful pointer to a stunning galactic treasure, the Sombrero Galaxy. In appearance it resembles a large, wide-brimmed Mexican hat and it also goes by the name of M104. The galaxy can be found about half a degree to the west of Spica but telescopes will be needed to see it. A 100mm telescope will show it as a smudge of light, 200mm will show the disc and bulge, while 250mm or above is needed to pick out the dust lane which gives it its characteristic appearance. It is hard to tell the shape of the galaxy because we are looking at it almost edge-on. Observing M104 is a great example of how different eye-pieces affect the view through a telescope. It is always tempting to pump up the magnification by using shorter-focal-length eyepieces to try and reveal more stars, but doing so means the image gets fainter. M104 illustrates how a compromise between image size and brightness needs to be found.

Heading a little further to the south-west from Spica is a relatively faint constellation called Corvus, which is made up from four moderately bright stars forming a four-sided shape. Its brightest star is called Gienah Corvi and it marks the north-west corner of the constellation.

Starting from Spica again, an imaginary line to the south will reach the third-brightest star in the sky, Alpha Centauri, which has a distinctly yellow colour to it. It is a member of the nearest star system to the Sun at around 4 light years. Halfway between the two, and on the southern side of the last two stars in the tail of Hydra, the snake, is the position of one of the nearest and brightest galaxies in the sky, known as the Southern Pinwheel, or M83. It is a stunning face-on spiral galaxy which is just beyond the limit of visibility to the naked eye, but binoculars will reveal the nucleus and a modest beginner's telescope will start to show some of its spiral structure.

Continuing on to the constellation of Centaurus, Alpha Centauri is its brightest star with the second-brightest, Beta Centauri, or Hadar, in the west. These two stars form a triangle with what is now the third-brightest star in the constellation, Epsilon Centauri, or Birdun. Taking the line between Hadar and Birdun and continuing on for the same distance again leads to a galaxy called Centaurus-A. Also known as NGC5128 this galaxy is close, at just over 14 million light years, yet even at this distance it is still necessary to use a decent pair of binoculars or telescope to spot it. Through larger telescopes it will appear as an elliptical galaxy with a dark dust lane superimposed against it, but radio telescopes reveal something quite surprising: it has two huge lobes of radiation extending out of its polar axis. It is thought the galaxy is the result of a merger event with the disruptive force of a black hole at its centre. Studies from the Spitzer Space Telescope have confirmed that its unique appearance is the result of an elliptical galaxy which is in the process of merging with a spiral galaxy.

Glancing a little to the south of Centaurus-A, it is easy to spot a faint smudge of light which is perhaps one of the most amazing sights in the night sky, Omega Centauri. It is known to be the largest and brightest globular star cluster orbiting the Milky Way and is home to several million stars. In reality it was once a dwarf elliptical galaxy that became captured by our own galaxy, with the discovery of a black hole in its core supporting this theory.

Arching through the southern part of Centaurus, running from east to west, is perhaps the most spectacular galaxy of them all, our own, the Milky Way. Measuring a staggering 100,000 light years from one side to the other, it is vaster than we can easily visualize, although it is fairly average in size compared to other galaxies. Our Solar System is around 30,000 light years away from the centre, which lies in the direction of Sagittarius over in the east at this time of year. The band of light we see as the Milky Way comes from the combined light of up to 400 billion stars that make up the galaxy, but scan along it and it is possible to pick out dark patches which are huge interstellar dust clouds that one day may form the next generation of stars.

Head back towards Alpha Centauri again, the bright yellow star inside the Milky Way, and just to its west is the constellation called the Southern Cross, or Crux. It does not only point towards the South Celestial Pole but also to a couple of faint smudges in the sky that could be mistaken for disconnected parts of the Milky Way. They are actually two of the satellite galaxies of the Milky Way, the Large and Small Magellanic Clouds, named after the explorer Ferdinand Magellan, who observed them on his voyage in 1519, although they were known about long before that. In reality they do not orbit the Milky Way but instead are galactic visitors, thought to have arrived here around 2 billion years ago.

The Large Magellanic Cloud lies in the constellation of Dorado, is easily visible to the naked eye and at 14,000 light years in diameter is about twice the size of the Small Magellanic Cloud found around 20 degrees to the west in the constellation of Tucana. The Large Magellanic Cloud shows evidence of high levels of star formation, hinting at its gravitational interaction with the Milky Way. The Tarantula Nebula is a fine example of one of these star formation regions and is just visible to the naked eye at the eastern end of the galaxy.

FOUR
The Invisible Universe

WHAT WAS THE LAST concert you went to? Whatever the kind of music, it is incredible to hear all the different sounds coming together to produce the songs that you were probably humming for several days after. Listen carefully to a rock band, say, and you can pick out the beat of the drum, the noises from the other individual instruments and the different vocal harmonies. Now, just imagine tuning in to the beat of the drum and nothing else: you would have no idea what the song actually sounds like. You have to tune in to all the elements before you get to hear the complete piece of music.

As with music, if we want to get a full picture of the Universe in all its glory we must tune in not only to visible light but to all the different types of radiation; otherwise we are missing out on an incredible amount of information. Our eyes have evolved to detect visible light, which is just a tiny portion of something called the electromagnetic (EM) spectrum. You are probably familiar with the 'colours of the rainbow' – red, orange, yellow, green, blue, indigo and violet – but extending either side is a much larger range of radiation that we are incapable of detecting with our eyes. Beyond the red end we find infra-red and radio waves; and beyond violet are ultra-violet, X-rays and gamma rays. To make that a bit clearer, see the illustration at the bottom of this page.

You can see an example of how different objects give off different types of radiation simply by looking up at the stars. There is an easy-to-recognize constellation called Leo which lies just north of the celestial equator. Leo can be seen from most parts of the world, although it will appear to be upside down from southerly latitudes. The brightest stars in the constellation are Regulus, marking the base of the lion's head, Denebola, marking its tail, and Algieba, which is the lion's mane. Look closely at the stars and it can be seen that their light is not the same: Denebola, shining white, is not too dissimilar to the blueish hue of Regulus, but both are

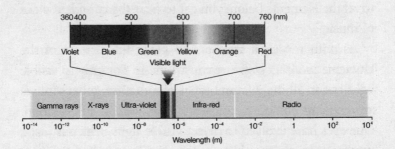

The Electromagnetic Spectrum

distinct from Algieba, which is orange in colour. These variations are the result of the distinctive temperatures of the stars and show nicely that stars emit radiation at different wavelengths in the visible-light range. Of course, they emit radiation in wavelengths other than visible light too, so to be sure we are capturing all available information we must study objects in all ways possible.

The real beauty of studying the Universe in all these various types of radiation is that you can reveal detail and information that otherwise would be lost to you. There is a challenge here though, since not all types of radiation make it to the surface of the Earth. Our atmosphere blocks much of the incoming radiation from space so, in a number of cases, we have to launch a telescope into orbit, beyond the filtering effects of our atmosphere.

Before we get too far into looking at how our view changes by tuning in to different types of radiation, it is worth briefly looking at how they actually vary. All types of radiation in the EM spectrum share the same wave-like properties and, for the purposes of this discussion, it is wavelength and frequency that interest us. Just like a wave in a lake, a wave of radiation has a crest which is followed by a trough before another crest arrives. The distance between two successive wave crests or troughs is called wavelength, and the number of waves that go past in a second determines frequency. More waves per second equates to a higher frequency and fewer waves to lower frequency. Scientists use these terms to describe the radiation's behaviour.

Considering the EM spectrum in its entirety, it starts with radio waves, which have the longest wavelength at thousands of metres (lower frequency), continuing through visible light to the other extreme, the shorter wavelength gamma rays at 100,000 millionths of a metre (higher frequency). (There is a mathematical relationship between wavelength and frequency but we do not need to worry ourselves about it now.)

The really useful thing about the EM spectrum is that different types of radiation can punch through different materials. If you have been outside on a clear dark night well away from street lights, then you may have seen the Milky Way stretching overhead and perhaps spotted some dark patches among the hundreds of glittering stars. These are huge dark dust clouds which lie between us and the more distant stars. They are so dense that they absorb visible light and block the distant starlight from our view. It turns out, though, that while these clouds block visible light, they do not block other types of radiation, so if you tune in to radio waves, for example, you can look right through them. This principle applies to clouds in our atmosphere too. Radio telescopes, which as their name suggests tune in to radio waves, can still work in the thickest of clouds, whereas optical telescopes are rendered useless.

The interaction of different parts of the EM spectrum with different materials allows astronomers to probe the deepest, darkest reaches of the Universe. This is not just of use to the professional astronomer though. A blight for the amateur astronomer is the ever-increasing level of artificial lighting in residential areas, but we can now use very special filters to cancel out most of this, effectively blocking out parts of the EM spectrum, leaving the visible light from the cosmos to shine through. We are already familiar with visible light and what we can see with it, so for now let us concentrate on the stuff we cannot see. Moving from visible light and heading to the right along the spectrum, we leave red light behind and enter the realm of infra-red radiation.

You already detect infra-red radiation (or heat, to use its common name) without even thinking about it. The warmth from a fire and even the heat from the Sun on a summer's day are examples of infra-red radiation. It was discovered back in 1800 by Sir William Herschel, a Hanover-born German astronomer who was trying to measure the temperature of each colour of the visible spectrum. He set about passing light from the Sun through a prism, which split the incoming white light into its component colours. He then placed a thermometer in each colour in turn and a further one at either end, beyond the visible light, to act as a control for his experiment. During this exercise, he noticed two things: that the temperature seemed to increase from the violet end to the red end and, more surprisingly, that the thermometer beyond the red end, where there was no 'light', seemed to register a higher temperature than in the red! Further experiments by Herschel showed that this new radiation, which he called 'calorific rays', acted just like the light he could see.

Following Herschel's discovery of what we now call infra-red radiation, it was found that water vapour would absorb it, a real nuisance for astronomers wishing to observe the sky in infra-red since 1 per cent of the Earth's atmosphere is made up of water. This means the greater part of incoming infra-red radiation is absorbed and only the smallest amount reaches us here on the surface. We get around this by placing telescopes high up in the atmosphere, above the majority of the water vapour – on top of high mountains, for example. NASA have taken this a step further with their airborne SOFIA facility, the Stratospheric Observatory for Infra-red Astronomy, which is a converted Boeing 747. Infra-red telescopes have even been launched high above the Earth's atmosphere, allowing them unimpeded opportunities to observe infra-red. The Infra-red Astronomical Satellite, or IRAS, is the first of its kind and was launched in 1983, becoming the first space observatory to survey the entire sky at infra-red wavelengths.

The measure of temperature, which effectively defines how much infra-red energy is being emitted, is actually a measure of the movement or vibration of atoms. If they are stationary then the temperature will be the lowest it can ever be, and is said to be at absolute zero. This means they will be emitting no infra-red energy, but as their movement increases the temperature goes up and so does the amount of energy being given out. The real benefit of infra-red studies is that they allow us to examine the temperature distribution of objects even if they are blocked from view by thick interstellar dust clouds.

We can study these interstellar clouds, or nebulae to use their correct term, in great detail. A fine example is the

Orion Nebula, found in the constellation of the same name that we looked at in the January sky. Just below the stars in the belt of Orion, for viewers in the northern hemisphere, or just above them for those in the southern, is a faint row of three 'stars' which depict the hunter's sword. From a dark location, look a little closer at the centre star and you can see that it is not a pinpoint of light but looks a little fuzzy. Try looking at it through binoculars or a telescope and you will discover an amazing sight: wispy strands of clouds surrounding what appear to be four faint stars in the middle. Infra-red studies show many more stars than the four easily visible ones, and we will look at how they are created out of these vast interstellar clouds in a later chapter.

Moving a little further along the EM spectrum we enter the world of radio waves and, in particular, a specific type of radio wave called microwave radiation. If you, like me, are a totally uninspired chef then you will know the real beauty of this lies in its ability to cook food. Inside those handy little ovens is a device that generates microwave radiation, which is in turn absorbed by the water, fat and sugar molecules in the food. As the molecules start to vibrate thanks to the extra zap of energy, they heat up, cooking your meal. Neat.

For astronomers, though, microwaves represent the most intense type of radio waves, but their first detection coming from the sky was really a matter of luck. In 1964, Arno Penzias and Robert Wilson were working at the Bell Laboratories in New Jersey and were experimenting with a special type of telescope that was designed to pick up microwaves bouncing off balloon satellites (a type of satellite which is spherical in shape and simply reflects communication signals back down to Earth, unlike conventional satellites, which receive signals and then retransmit them).

In order to be able to pick up the faint reflected signals, they had to remove as much interference as possible, which they managed, except for a faint yet persistent signal. It remained pretty constant in whichever direction they pointed the horn-shaped telescope and, suspecting pigeon droppings, they set about giving this a thorough clean. The signal continued and they eventually concluded that it must be coming from beyond the Earth, from deep space! Penzias and Wilson had in fact detected the remnant radiation from the birth of the Universe, the Big Bang, which occurred 13.7 billion years ago.

This is a good opportunity to consider a rather strange concept in astronomy. Light travels at almost 300,000km per second and in our everyday lives this is of no real consequence. In astronomy, though, it is incredibly useful to know this fact. Because of the vast distances across the cosmos it takes time for light to get from one place to another; to the Earth from the Moon it takes just over a second, from the Sun 8.3 minutes, from the nearest star around four years, and from the nearest neighbouring major galaxy 2.3 million years. This means that by looking at distant objects, we are actually looking back in time. The latest studies from the space-based microwave telescopes, known as the Wilkinson Microwave Anisotropy Probe (WMAP), have looked at what we now call Cosmic

Background Radiation in unprecedented detail and determined that its source is so far away that it gives us a view of the Universe as it was just 400,000 years after the Big Bang.

If you look at the night sky with your own eyes you will see it as visible light. You will see individual specks of light, most of which are stars, but some of these tiny pinpricks will be planets; others that look a little fuzzy will be clusters of stars or even distant galaxies like the Andromeda Galaxy in the northern sky. In between the stars, it will seem dark and black, except of course if you look towards the inside of our own galaxy, the Milky Way. But if you could look at the sky as microwave radiation then you would not see many, if any, of the stars you are familiar with; instead, you would see the Cosmic Background Radiation covering the entire sky. It would be a sky full of light with a few darker patches here and there. It is believed that the slight variations in the intensity of the radiation ultimately led to the evolution of the vast galaxy clusters that we see in the Universe today.

It was some thirty years before Penzias and Wilson revealed microwave radiation in the sky that a telecommunications engineer called Karl Jansky discovered the more general form of radio waves from astronomical objects. He was investigating interference detected on transatlantic voice transmissions and was using a directional antenna that allowed him to tell where a signal was coming from. He picked up some interference that seemed to be peaking every 23 hours and 56 minutes which, in consultation with an astrophysicist, he concluded must have been astronomical in origin because it took the Earth that long to rotate on its axis. By comparing his results with maps of the night sky, he saw it could only have been one thing: he had been picking up radio signals coming from the centre of our galaxy.

The real beauty of the great proportion of the radiation at the radio end of the spectrum is that it nearly all passes through the atmosphere, so it can be observed from the ground – and in broad daylight. This means that radio astronomy studies of the Universe can continue around the clock almost regardless of weather conditions, except perhaps during some thunderstorms.

The patch of sky where Jansky had first detected astronomical radio waves lies in the direction of Sagittarius and the galactic centre of the Milky Way. To the naked eye this area of sky does not look like anything special, although it does seem pretty much packed with stars if you study it with a telescope. The main reason for this, though, is that most of the light coming from the centre of our galaxy is blocked by interstellar gas clouds but, interestingly, it is these gas clouds that contribute to the radio waves detected by Jansky in the early 1930s.

Today's professional radio telescopes are a little larger than the directional kind used by Jansky. They are usually shaped like a huge dish and in size tend to dwarf all other types of telescope. One factor that determines the diameter of any telescope is how much detail it can see; in other words, its resolution. The bigger it is, the finer the level of detail it can detect. The wavelength of radiation being studied also has a bearing on this and a longer wavelength needs a bigger telescope to see the same level of detail that a shorter-wavelength telescope can see. This is why radio telescopes are so huge. That said, it is possible for amateur radio astronomy set-ups – and I've even seen a pretty rudimentary radio telescope made out of a bin lid and a few basic electronic components – to detect radiation from Jupiter and the Sun.

Another great way to try radio astronomy for yourself is from the warmth and comfort of your car during a meteor shower. The Lyrids are a good example of a meteor shower in the April sky and peak around the 20th/21st of the month. Whether or not a good show is seen depends on the amount of meteoric material the Earth encounters, the orientation of the Earth at peak activity, cloud, light pollution and even the Moon. The trick is to tune your car radio to a commercial FM station that you cannot normally pick up, ideally one that is around 1000km away. For now you will just hear the hiss of noise, but as meteors zip in through the atmosphere radio waves will bounce off their trail, allowing you to hear the distant station for a brief moment. You might also hear pops and whistles as the meteors arrive.

That is it for the longer-wavelength end of the electro-magnetic spectrum; moving back to the other end of the visible portion we come to violet light. Beyond violet light is ultra-violet, or UV, radiation and its discovery was, like many scientific finds, a matter of chance. By the start of the nineteenth century infra-red radiation, and its warming effect, was known about, so in 1801 a German physicist called Johann Ritter set about looking for an equivalent radiation that might have a cooling effect. He started looking in the dark regions beyond violet light and, while he did not find any 'cooling' light, he did notice that a white chemical called silver chloride turned dark when placed beyond violet light. He had discovered UV radiation, although he first called it 'chemical rays' as a result of its observed ability to produce chemical changes.

Many birds and insects can see in UV but humans are blind to it, although we can 'detect' it from the way the Sun burns our skin. The ozone layer in our atmosphere blocks around 95 per cent of the incoming solar and astronomical UV radiation, so astronomers wishing to study it must once again send telescopes up into space. The view through telescopes that can see in UV, e.g. the Hubble Space Telescope, is quite different since most stars are in the middle years of their lives and surprisingly are relatively cool, while ultraviolet radiation is the mark of hotter objects like stars at the beginning or end of their lives. Unlike radio waves, which can pierce galactic dust clouds, UV is blocked by them, so it is very difficult, but not impossible, to study the stars of the Milky Way in UV. It can be used effectively to study the chemical make-up of the interstellar clouds and also to probe distant galaxies and learn about their evolution.

Moving beyond UV poses a challenge for astronomers because the radiation has shorter wavelengths so is much more energetic, which means it blasts through most materials if pointed straight at them. This property is used very successfully in medicine, which exploits the ability of X-rays to see inside humans without having to subject them to an operation. This does mean, though, that when researchers use a conventional telescope designed to capture incoming light, the X-ray radiation has so much energy that it will fly straight through the mirror. The solution is to tilt the mirror so that the light strikes it at a shallower angle.

Because X-rays are blocked by the atmosphere, the high-energy telescopes used to study them must be placed into Earth orbit. The first source of X-rays discovered beyond the Sun is known as Scorpius X-1, which has been studied and found to be a type of star known as a neutron star. Nothing amazing about that, but what does make it special is that it is part of a binary star system too. Neutron stars have an incredibly strong gravitational pull and can literally rip material off a companion star. This is the case with Scorpius X-1, and the material builds up into a disc which is accelerated to incredibly high speeds, causing it to start to emit X-rays, which is the radiation we can detect from 9000 light years away (the distance light can travel in 9000 years).

Beyond X-rays is the final piece of the EM spectrum and it represents some of the most energetic and violent events in the Universe: gamma radiation. It had been known about and observed in natural processes here on Earth for many years – for example, during radioactive decay – long before it was discovered in deep space. The existence of this most energetic form of radiation had been predicted during extreme events such as the supernovae that mark the death of a supermassive star, but it was not until the late 1960s that it was actually detected. Satellites from the Vela military satellite group were searching for bursts of gamma rays on the ground from the detonation of nuclear weapons, but they also picked up flashes of gamma rays from deep space. Careful study later showed that these brief bursts of gamma radiation lasted for only a fraction of a second and came from totally unconnected parts of the sky, quite randomly.

As I've touched on before, these brief events are known as gamma ray bursters (GRBs) and when a star explodes it sends out a burst of energy so great that in just a few seconds it produces more energy than our Sun over its entire 10-billion-year lifetime! Unfortunately GRBs are incredibly hard to study because they are so short-lived. However, following the initial burst, an afterglow is often present that emits radiation in longer but fading wavelengths. If astronomers can respond swiftly to a GRB event, it is possible for them to detect the afterglow and identify the origin.

GRBs are a fitting way to end this section as the only way to understand them is by studying them over a range of different wavelengths. Without this approach, they would remain a mystery. As with the rock band at the start of the chapter, it is only by opening our senses to the whole EM spectrum that we can ever hope to get a full picture of the cosmos.

April: Northern Hemisphere Sky

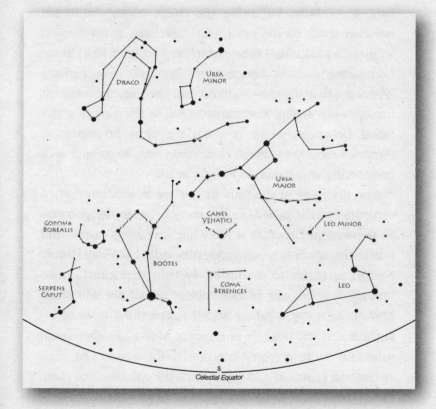

In the previous section I referred to a constellation called Leo and it is a great place to start the April guide. To find it, look just to the north of the celestial equator and a little to the west to find a pattern of stars that resembles a backwards question mark. This marks the head of Leo and at the southern end of the grouping is the brightest star in the constellation, Regulus, which is a blue-white giant star 78 light years away. Following the shape of the backward question mark to the north, the next star is Eta Leonis, followed by Gamma Leonis, which is 125 light years away. Comparing Gamma Leonis with Regulus, it is perhaps surprising that Regulus is almost half the distance away. At the opposite end of the constellation to the east is a star called Denebola, which is about the same brightness as Gamma Leonis yet is much closer than even Regulus at well under half the distance at 34 light years.

The distances to the stars in Leo are insignificant when compared to the incredible distances between the galaxies. To the west of Denebola is the white star Theta Leonis and just to the south is a pair of galaxies called M65 and M66. Neither is visible to the naked eye but they can be seen through a good pair of binoculars. Telescopes will show M65 to be a spiral galaxy which is presented to us at an angle and is the brighter of the two; M66 is another spiral galaxy but more face-on. Compared to the stars in Leo, such as Gamma Leonis at 125 light years, the galaxies beat them hands down at a staggering distance of around 35 million light years.

Off to the south-east of Denebola is the constellation of

Virgo and scattered over its northern boundary is the Virgo
Cluster of galaxies. The cluster is thought to span around 15
million light years of space, which is not much larger than
our own Local Cluster of galaxies, although it has significantly more members. Red shift measurements taken by
studying the spectrum of some of the Virgo members show
how fast they are travelling away from us, and from that it is
possible to calculate that the centre of the cluster is about 54
million light years away.

Just to the north-east of Virgo and its cluster of galaxies is
the unmistakably bright orange star Arcturus in Boötes,
which is a strong contrast to the blue-white colour of
Regulus in Leo. Not only is it the brightest star in the constellation but it is also the brightest star in the northern
hemisphere of the sky. The name Arcturus means 'bear
watcher', which has its origins in the fact that it follows Ursa
Major, the Great Bear, as it circumnavigates the Pole Star. At
a distance of 37 light years it is the nearest giant star to us,
giving a great opportunity to study the evolution of these
stellar monsters. Moving to the east of Arcturus takes us to
Zeta Boötis, which is a binary star system 180 light years
away. The stars of this system orbit around a common
centre of gravity but along unusually elliptical orbits.
Extending the line between Arcturus and Zeta Boötis points
to the brightest star in the constellation Serpens, right next
to Virgo. This star is called Alpha Serpentis and to its south-
west is a tiny fuzzy-looking blob just visible to the naked
eye. This is the globular cluster called M5 and it lies at an
estimated 24,500 light years away. As with many globulars,

it is thought to be home to a huge number of stars, maybe as many as half a million, and it is their combined light which gives it the appearance of a fuzzy blob.

A little to the north-west of M5 is the location of the largest known galaxy, with the most uninspiring name of IC1101. Regardless of its monstrous size – it is over fifty times bigger than our own galaxy – at a distance of 1.07 billion light years it appears only as a faint smudge in the sky so is beyond the range of most amateur telescopes.

Skipping back due west from Arcturus is Diadem, the brightest star in Coma Berenices, and it lies at a distance of 47 light years. Just 1 degree to the north-east of Diadem (remember, you can estimate this distance as it is the same as a finger extended up at arm's length) is the globular cluster known as M53, and a degree to the south-east of that is NGC5053. Although a little fainter, NGC5053 is still visible through amateur telescopes but, of the two, M53 is by far the easier to find. The distance to M53 is 58,000 light years, which means it is among the most distant of the Milky Way's globular clusters, while NGC5053 is a little closer at 53,500 light years.

Hopping back due west to Leo and then north is a rather less conspicuous constellation representing the lesser lion, Leo Minor. Even the brightest star in the constellation struggles to shine brighter than 4th magnitude and the other stars, stretching out to the west in a rather haphazard line, are fainter still. To the north of Leo Minor is the constellation of Ursa Major, containing the famous arrangement of the stars which resemble a Plough or Big Dipper.

The stars in the Plough are easily seen and very familiar to northern hemisphere observers, but those in the rest of Ursa Major are less prominent. The head of the bear stretches out to the west, with its front leg to the south, and the rear leg extends south from Phecda, the star at the south-east corner of the pan of the Plough. There are around 200 notable stars in the constellation and their distances vary wildly. Of those easy to see with the naked eye the nearest star, Alula Australis, is found just to the south of Alula Borealis, which is the star at the bottom of Ursa Major's rear leg and a mere 27.3 light years away. The most distant star easily visible with the naked eye is 83 Ursae Majoris, at a distance of 549 light years and found between and slightly to the north of Alkaid and Mizar, the two end stars in the bear's tail. This distance, though, is nothing compared to the most distant star in the constellation, T Ursae Majoris, at an impressive 5930 light years and located between and slightly to the north of Alioth and Megrez, the two most westerly stars in the tail.

The two stars at the western end of the pan of the Plough, known as the Pointers, point directly to a 2nd magnitude star called Polaris, the North Star. It lies 430 light years away and marks one end of Ursa Minor, the Lesser Bear, which looks just like a small version of the Plough. Between the north of the tail of Ursa Major and the pan of Ursa Minor is the end of the constellation Draco, which is one of the largest in the northern hemisphere sky. From here, it curves round behind Ursa Minor and finishes to the north of Hercules in the south-east.

April: Southern Hemisphere Sky

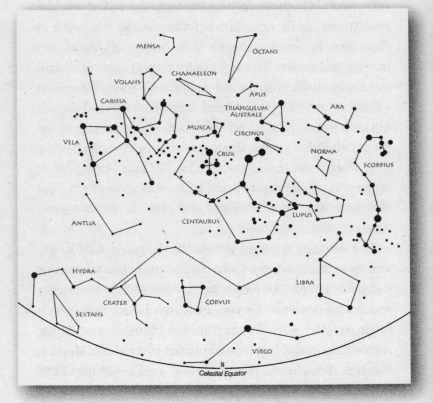

Spica in Virgo offers a great starting place for our April guide to the southern hemisphere sky as it is just south of the celestial equator and in a fairly sparse area of sky. Careful study of the light from Spica reveals that it varies over a regular period of about four days. The reason is that it is not a single star but in reality a binary star system with the two stars in orbit around each other at only 18 million kilometres apart, which is nearer than the Sun and Mercury. The close proximity of the stars means they complete an orbit of each other in just over four days, leading to the regular dip in brightness as one of the stars eclipses the other. Just to the south-east of Spica is a constellation of much fainter stars, called Corvus, whose stars are a little brighter than 3rd magnitude, compared to the much stronger Spica at around 1st magnitude.

The stars in Corvus are arranged in a distorted square shape and are all broadly the same brightness as you look at them in the sky, yet their distances vary. Starting at the star nearest Spica then moving around to the south are Algorab at 88 light years, Kraz at 140 light years, Minkar at 303 light years and finally Gienah Corvi, the brightest star in the constellation, at 165 light years. Just to the west of Kraz at the south-west corner of Corvus is Cauda Hydrae, the second-brightest star in Hydra, the snake. Hydra is the largest constellation in the sky and winds across the top of Corvus, Crater and Sextans to the east before stopping at Cancer in the northern hemisphere sky.

Directly to the south of Cauda Hydrae is the prominent constellation Centaurus, with its brightest star, Rigil

Kentaurus, or Alpha Centauri as it is more commonly known. Any uncertainty in identifying this star can be resolved due to its proximity to Hadar just a few degrees to the east. Alpha Centauri is a multiple star system of three stars: Alpha Centauri A and B, which orbit each other over an eighty-year period, and Proxima Centauri, which is a red dwarf star. At a distance of 2.2 trillion kilometres from the other two it is uncertain whether Proxima Centauri is gravitationally bound to them or just a visiting star on its way through the system. Either way, what is certain is that it is the closest star to our Solar System at 4.22 light years, which is 40 trillion kilometres. The distance to Proxima Centauri is easily determined by measuring its parallax, which is its apparent shift caused by the movement of the Earth around its orbit.

To the east of Alpha Centauri and Hadar are the stars of the Southern Cross, or Crux, which looks like a cross with its long axis aligned north–south. The most southerly star of the cross, Acrux, is also the brightest in the constellation and lies 320 light years from us. It is a multiple star system where two of its stars can be easily separated in small telescopes, but the brighter of the two is also a spectroscopic binary which means its hidden companion is only visible through studies of the spectrum of the star. The most westerly star of the cross is called Becrux, and lying to its south-west is perhaps one of the finest star clusters in the sky, the Jewel Box. It earned its name from Sir John Herschel, who when describing what he saw through his telescope said it looked like 'a casket of variously coloured precious stones'. A telescope is not essential to appreciate this object though, as even binoculars will reveal its glittering collection of stars.

A little to the east of Acrux is Lambda Centauri, in Centaurus, which as a constellation straddles the Southern Cross. The star lies 410 light years from us, which means the light we can see today left around the time Galileo first turned a telescope on the sky! To the south-west of Lambda lies a cluster of stars embedded in beautiful nebulosity, visible to the naked eye and at a distance from us of 6000 light years. In photographs the nebula appears red in colour, which is a trademark of an emission nebula. In these interstellar nurseries the energy from the hot young stars causes the atoms to glow, giving off their own light at a very specific wavelength. The nebula, which is often called the Running Chicken Nebula because it supposedly resembles a chicken mid-dash, is only visible from a dark site and often only when employing a technique called averted vision. This observational skill simply means looking slightly to one side of a faint object, which allows the light to fall on a more sensitive part of the eye and make it appear brighter.

Directly to the south of Crux and its brightest star, Acrux (which is a shortened version of Alpha Crucis), is one of the smallest constellations in the sky, Musca. It traces out the shape of another distorted square, with its brightest star, Myla, which is the second-most northerly star in the constellation. In the south-west corner is Delta Muscae, which is distinctly orange in colour, and to its north is NGC4833, a lovely example of a globular cluster. It is just beyond the limit of visibility to the naked eye but an easy target for binoculars even though it is 21,200 light years away. NGC4833 is, like most globulars, composed of old stars and is one of hundreds of clusters in a halo around our galaxy. It was studying the distance and distribution of objects like this that led astronomers to gain insight into the size and shape of our galaxy.

The Southern Pleiades is an open or galactic cluster which can be seen to the north-east of Musca by the naked eye. Unlike the globular clusters such as NGC4833, open clusters are composed of young hot stars and are around 50 million years old. The name comes from its appearance, which is similar to the Pleiades star cluster seen in the northern hemisphere sky in the constellation of Taurus.

The real showcase of the southern sky in April is the Milky Way itself, an unmistakable glow of light stretching across the sky from east to west. It is easy to see why understanding the shape and size of our galaxy was a tricky business − after all, we live inside it and have no way of assessing it from afar. The study of globular clusters started to unlock its secrets, but following on from that it was mapping the interstellar dust clouds like the Running Chicken Nebula in Centaurus that gave us the final answers. By their nature, the stellar nurseries are most common in the spiral arms, which meant studying them enabled a map to be built up of the structure and scale of our galaxy, with its centre about 30,000 light years away in the direction of Sagittarius in the western sky.

It is easy to see how our ancestors thought all objects in

the night sky were stuck onto giant crystal spheres surrounding the Earth. Observations of real objects over many many years, some of which have been covered here, have slowly changed our view of the Universe and given us a real three-dimensional impression of our place within it.

FIVE

The Scale of the Universe

EVERYONE KNOWS THE Universe is a big place but just how big is it? It is very easy to rattle off distances in the Universe, from the Earth to the Moon at around a quarter of a million kilometres, to the Sun at 150 million kilometres, all the way out to the most distant object ever seen at 124,000 million trillion kilometres! Yet we cannot use a giant tape measure to work these out so how have the immense distances in space been calculated? The answer is, simply with a dash of human ingenuity.

As we've seen, the Greek mathematician and astronomer Eratosthenes once used shadows cast by the Sun to calculate, with some accuracy, the circumference of the Earth. It was this discovery that laid the foundations to allow mathematicians and scientists through the ages to gain an understanding of the distances in space. Knowing the circumference of the Earth, Eratosthenes then successfully calculated the distance to the Sun, but little is known of the method he used. He estimated it to be 'of stadia myriads 400 and 80,000' but it is not known whether this means 4,080,000 stadia or 804,000,000 stadia. It all hinges on the application of the phrase 'stadia myriads', where a stadion is considered to be around 185m and myriad refers to a multiplier of 10,000. It could mean either 400 myriad plus a further 80,000 or simply 80,400 myriad and, interestingly, if we understand it to be the latter, then it translates to 149 million kilometres to the Sun. This figure is impressively close to our current average figure of just under 150 million kilometres.

One of the later methods used for calculating the distance between the Earth and Sun was demonstrated by Jeremiah Horrocks in 1639, when he observed a transit of Venus. This event occurs when the planet Venus passes directly between the Earth and Sun making it visible against the bright solar disc. Using projection techniques to observe the event safely through a telescope, Horrocks and a colleague timed the exact start and end time of the transit from their separate observation points. Thanks to Johannes Kepler's Laws of Planetary Motion published earlier that century, they also knew the relative distances of the planets based on how long they took to go around the Sun. Using these measurements for the Earth and Venus and the apparent shift in the position of Venus from the different observation points, Horrocks was able to calculate the distance to the Sun as a multiple of the radius of the Earth. The answer he came up with was 13,750 Earth-radii which, using the modern figure of 6378.1km, gives a figure of just under 88 million kilometres.

Although the figure Horrocks suggested is significantly lower than the currently accepted figure of just under 150 million kilometres, the principle of the experiment was sound; only its execution was wrong. Later in the seventeenth century the approach was improved upon by James Gregory and a more correct figure obtained. More recently, radar measurements of the position of Venus have enabled an exact distance of Venus to be determined, which finally allowed the distance between the Earth and Sun to be accurately calculated. The actual distance between the two varies throughout the year due to the elliptical nature of the Earth's orbit, but the average distance is used as a unit of measurement within our Solar System and, called the astronomical unit, has a value of 149,597,871km.

Other than the astronomical unit, there is another popularly used measure in astronomy, which I've already mentioned: a light year equates to the distance light can travel in one year in a vacuum. The value of one light year is equal to just under 10 trillion kilometres. The first time it was demonstrated that light travelled at a finite speed was in 1676, by the Danish astronomer Ole Rømer, but it was sixty years earlier, in 1616, that the foundations were laid for his historic experiment. The determination of longitude at sea was a major problem, but Galileo suggested that the moons of Jupiter could be used as a celestial clock and, by using them to determine the local time, longitude could be calculated. For this to work, accurate predictions of the movement of the four moons of Jupiter were needed. Unfortunately Galileo's idea did not work at sea due to the movement of a ship, but it proved to be a brilliant method on land.

The Italian astronomer Giovanni Cassini worked hard to improve the predictions of the moons of Jupiter and to that end needed some accurate observations and timings. He worked with another astronomer, the Frenchman Jean Picard, and his assistant, Ole Rømer, who between them logged hundreds of observations. From these observations, Rømer noticed that the time period between successive events of Jupiter's innermost moon, Io, seemed shorter when Earth was moving towards Jupiter and longer when it was receding. His conclusion was that it must take a finite period of time for light to travel the extra distance across the orbit of the Earth and he determined this delay to be around 22 minutes. The Dutch astronomer and mathematician Christiaan Huygens took this figure and, combining it with an estimate for the diameter of the Earth's orbit, arrived at a figure for the speed of light of 220,000km per second, not bad compared to our current figure of about 300,000km per second.

Measurements within our Solar System tend to be expressed in the astronomical unit (Au), and radar can be used to measure distances to the nearer planets, so with mathematics it is possible to calculate the distances to other planets with astonishing accuracy. Measurements beyond the Solar System are generally described in light years, but one other measure is also used, the parsec, and this has its origins in the Earth's orbit around the Sun.

The parsec is equal to 3.26 light years or approximately

31 trillion kilometres and its name comes from its definition, in which the distance corresponds to a PARallax of one arc-SECond. This is not as complicated as it sounds. Consider the Earth in its yearly orbit around the Sun, the diameter of which orbit is about 300 million kilometres. As the Earth moves on its orbit it changes its position in space, so much so that nearby stars seem to shift due only to the movement of the Earth. This apparent shift is known as parallax and you can demonstrate it for yourself now. Extend your arm in front of you and point one finger up. Shut one eye and notice where your finger is in relation to a background object. Now open that eye and shut the other, and without moving your finger it will seem to have shifted its position against the more distant background objects. If you were to measure the angular shift of your finger (assuming you are at the centre of a 360 degree circle, your finger will have moved by just a few degrees) and also measure the distance between your eyes, you could calculate the length of your arm. In the case of one parsec, the shift in position or angular shift is one arc-second, which is equal to 1/3600th of a degree (one degree is the area of sky equivalent to the width of a finger held at arm's length).

The same approach is used in astronomy to measure the parallax of relatively close stars, but instead of using the distance between two eyes, measurements are taken six months apart when the Earth is at the extremes of its orbit. We know the distance to the Sun and therefore the diameter of the Earth's orbit, so by measuring the apparent shift of a star we can work out how far away it is. The parallax method of measuring distances in space is used by the Hipparchus spacecraft, which can measure stellar parallax at about 1500 light years. To put this in context, the parallax of the nearest star, Proxima Centauri, is the same angle that a large coin would display at a distance of about 5 kilometres.

A parsec then is a measure of distance just like metres, kilometres and light years, where one parsec is the distance a star will be from Earth if it displays a parallax shift of one arc-second (the full moon, for example, measures 1800 arc-seconds in diameter). Like many metric measures, there are further multiples of this, such as the kiloparsec (1000 parsecs) and the megaparsec (1 million parsecs).

As technology improves it will be possible to measure even smaller parallax shifts, but to measure the vast distances to the galaxies, another approach is needed. Before looking at how this can be done, it is useful to put into context the sheer vastness of space. The nearest major

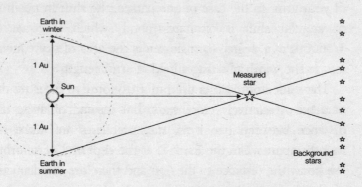

The Parallax Method

galaxy to our own, the Andromeda Galaxy, is 2.3 million light years away (approximately 770 kiloparsecs). That means that if you could travel at the speed of light, which is almost 300,000km per second, it would take you 2.3 million years to get there. Even the Helios 2 space probe, which holds the record for the fastest manmade object at an incredible 252,800km per hour, would take over 10 billion years to get there. The most distant object discovered so far is a galaxy called UDFj-39546284, which lies at a distance of 13.2 billion light years.

Measuring the huge distances in intergalactic space requires a special technique, and some of the galaxies close enough for individual stars to be detected hold the secret to this. Back in the first half of the 1900s Henrietta Leavitt was studying variable stars in our satellite galaxies, the Large and Small Magellanic Clouds. She spotted around fifty stars that seemed to share a special relationship: they were brighter and had longer periods of variability than those that were fainter. Making the assumption that they were all at roughly the same distance she realized the differences in brightness in the sky or apparent magnitude would mean an equal difference in real luminosity or absolute magnitude (see Chapter 9 for a full description of this). Studying the results in detail, she discovered there was a distinct relationship between the time it took the stars to change in brightness and their actual brightness, the so-called period–luminosity relationship. The principle is simple: if you can see two of these so-called Cepheid Variable stars in the sky with the same period of variability but one is brighter than the other, then the fainter one must be further away.

Leavitt's discovery was of profound importance and turned out to be key in understanding the scale of the Universe. The Danish astronomer Ejnar Hertzsprung was the first to realize the significance of this. If one of these Cepheid Variable stars was identified from its light curve (a graph showing the way its brightness changes with time), a quick measure of the brightness with which it appeared in the sky would allow its distance to be calculated. The American astronomer Harlow Shapley was the first to apply this technique to measure distances in space and was able to use it along with other observations to calculate the size of our own galaxy, the Milky Way, and even to calculate our approximate position within it. The next leap in understanding came in 1924, when Edwin Hubble studied the Andromeda Galaxy and the spiral galaxy in the constellation of Triangulum and identified Cepheid Variables using the 2.5m telescope on Mount Wilson. From this, he calculated their distances as 900,000 light years and 850,000 light years respectively, and although we now know they are somewhat further away than this, it did conclusively show them to be objects well outside our own galaxy.

Cepheid Variables have been successfully used to measure distances in the Universe many times now, but this technique is limited to galaxies where individual stars can be discerned. A similar approach can be applied to the type 1a supernova, which is a white dwarf star that slowly collects matter from a neighbouring star. Over time, the increasing pressure from the matter slowly heats the core of the white dwarf star until a runaway fusion process starts, rapidly converting the star's material into heavier elements. As this continues, the temperature increases violently until the star rips itself apart. This brings with it a dramatic increase in brightness that can be seen across the vast distances of space.

Because the fundamental process that causes the type 1a supernova is dependent on a certain critical mass, the brightness that accompanies the final explosion is always the same. Not only is the same brightness observed but so too are characteristic features in the spectrum. With these reliable properties it is possible to identify a type 1a supernova and, just as with a Cepheid Variable, comparing its apparent brightness with how bright it really is can determine its distance. This method can be used over much greater distances than the Cepheid Variables approach.

To understand the final method for determining distances in space it is necessary to understand a little about the characteristics of light. If you put white light through a prism or a device called a diffraction grating, then the white light is split into its component parts and you see all the colours of the rainbow. The reason this happens is because the different wavelengths are bent or diffracted by varying amounts, so the individual coloured light is spread out. Extending further out either side would be the other types of radiation that were discussed in Chapter 4. Using specialist devices called spectroscopes, astronomers can look at the light coming from stars and separate it into its component parts, revealing its spectrum. When the German optician Joseph von Fraunhofer studied the solar spectrum in the first half of the nineteenth century he noticed strange dark lines superimposed on the rainbow of colours, but it was not until 1859 that the German chemist Robert Bunsen discovered their origins. He realized that tiny particles inside the atom, called electrons, were absorbing very specific wavelengths of light and stopping them from coming through to the observer. In the case of the solar spectrum, the electrons present in certain gases in the Sun, but also in the Earth's atmosphere, were responsible for absorbing some of the incoming light. This effect is of great use to astronomers as different gases produce specific patterns of lines, called absorption lines. They appear in very particular positions, and identifying these lines in the spectrum of a distant object allows its chemical make-up to be determined.

The absorption lines can be seen in the spectrum of stars and galaxies alike, and it is the latter that allows us to measure the vast distances to the edge of the observable universe. By studying the position of absorption lines in the spectrum of a star in 1848, the French physicist Hippolyte Fizeau noticed the expected pattern of lines had shifted very slightly from where they should have been and he put this down to a phenomenon known as the Doppler effect. The effect is named after the Austrian mathematician and physicist Christian Doppler, who explained that sound or light emanating from a moving object would change its frequency as the object moved. A popular example of the effect is a moving emergency vehicle: as it approaches, the pitch of its siren seems to go up, and as it moves away, the pitch goes down. The reality is that sound waves emitted by the siren are 'squashed up' as the vehicle approaches you, and you receive a higher-frequency noise (the pitch goes up), and the waves are stretched out as it recedes, leading to a lower frequency being received (the pitch goes down).

Just as Fizeau discovered, the effect is not restricted to sound waves but is seen in light too. It manifests itself in light as a shift in position of the absorption lines against the background spectrum. Vesto Slipher noticed in 1912 that most spiral nebulae seemed to show quite a significant jump in the position of the lines, all of which seemed to shift towards the red end of the spectrum. From measuring the apparent red shift it is possible to calculate the speed at which the object is moving away from the observer. Edwin Hubble took this one step further and realized there was a mathematical relationship between the speed the galaxies were moving away at and how far away they were. He had found the final piece of the mathematical puzzle (now called Hubble's Law) in measuring the incredible distances in intergalactic space, and with it the vast scale of the Universe was finally revealed.

Interestingly, the relationship between speed of movement and distance is related to something that goes back nearly 14 billion years, the Big Bang. Before this explosive event, the entire Universe, everything, was compressed into a tiny point called a singularity, and after the Big Bang the whole Universe, space included, was catapulted into violent expansion. It is important to realize that it is not the galaxies themselves that are shooting out in all directions but space itself, with the galaxies being dragged along with it. The red shift phenomenon is seen because space itself has grown since the light left its origin and, with the intervening expansion of space, the spectrum has been stretched, shifting the absorption lines.

It is difficult to really visualize the enormous dimensions of the Universe so it is sometimes useful to look at scaled-down representations, but even then to consider the entire Universe on one scale is almost impossible. For example, if the Earth were represented as a grain of sand, with the Sun about one metre away, then Proxima Centauri, the nearest star, would be 354km away, and the nearest galaxy just over 2 million kilometres away (that is, roughly six times the distance to the Moon).

It is not just the distances in space that are hard to imagine; the size of objects varies enormously too. The structure, evolution and even appearance of the Universe are dictated by the very small, so in considering the size of objects we should start at the atom. In the same way that it is hard to imagine the size of the very big, it is hard to imagine the size of the very small and atoms definitely come into this category. If an atom were enlarged to the size of a football, then in comparison a small coin would become the size of the Earth! There are even smaller components making up the atom: the neutrons and protons in the nucleus and the electrons in orbit around it. While this may seem the stuff of science fiction – after all, no one has actually 'seen' an atom – it is their influence that determines stellar evolution, the crushing power inside black holes and even the features in spectra that enable us to probe the nature of distant objects.

A little easier to visualize are the bits we can actually see, like the particles inside the glowing clouds called nebulae where stars form, through to the dust particles making up the rings of Saturn. The most common, comprising the interstellar dust that pervades the majority of galaxies, are tiny fractions, often millionths, of a millimetre in size. Moving up a little on the scale takes us to the average pieces of interplanetary dust, which we often see 'burning up' when they fall through our atmosphere as the streaks of light known as meteors. They range from the size of a grain of sand up to a few millimetres. This is in contrast to the size of the particles in Saturn's rings, which are mostly water ice, and vary from a centimetre up to several metres.

It is not just tiny pieces of dust and small chunks of ice that join the planets in the Solar System. The bodies making up the asteroids and comet nuclei are a little larger but they differ in composition, comets generally being more icy than rocky. The smallest of the asteroids and cometary nuclei are tens of metres across, while the largest asteroid, Ceres, measures 975km.

Ceres is actually considered a dwarf planet like Pluto, whose diameter is 2306km, making it the second-largest of the dwarf planets after Eris, which also goes round the Sun beyond the orbit of Neptune. Dwarf planets differ from planets because they have not 'cleared their neighbouring region of other objects' and it is because of its relatively large companion, Charon, that Pluto was demoted from its planetary status. The remaining 'major' planets vary in size considerably, from the smallest, Mercury, at 4879km across, to the mighty Jupiter at 142,984km. In fact, Jupiter is so large that you could fit all the other planets inside it and still have room to spare.

Surprisingly, even some stars are of comparable size to the planets, such as the red dwarf Proxima Centauri, which is thought to be around 200,000km in diameter (compared to Jupiter's 142,984km). There are even some stars like Sirius B, the companion star of the much brighter star Sirius, at just 12,000km, that are smaller than the Earth. It belongs to a class of star known as white dwarfs, which are remnants of massive stars that have blown most of their layers off into space, revealing the cool and dying core. There is a giant leap in magnitude between the smallest stars and the largest ones, with VY Canis Majoris measuring a staggering 2.7 billion kilometres. It is so large that you could fit just over 218,000 planet Earths across its diameter!

VY Canis Majoris is around 2000 times the size of the Sun, and if you were to put it in the Sun's position then it would be so large it would swallow up the orbits of Mercury, Venus, Earth, Mars and Jupiter, but stop just short of the orbit of Saturn. There is no doubt that VY Canis Majoris is a monster among stars yet even it is dwarfed by the Solar System. Defining the size of the Solar System is more about identifying its edge but this is no mean feat. The edge can only really be defined by the point at which it is difficult to differentiate between the influence of the Sun and the influence of the nearest star, or the interstellar medium. This distance is known as the heliopause and it is expected to lie somewhere between 8 billion and 14 billion kilometres from the Sun. In the next five to ten years Voyagers 1 and 2 are expected to confirm its exact location as they become the first manmade spacecraft to leave the Solar System.

The stars and their systems of planets (over 700 planets have been discovered around other stars) make up the vast proportion of objects found inside galaxies. Our own galaxy, the Milky Way, measures around 100,000 light years from side to side, which is on a completely different scale to the familiar size of the planets, but as galaxies go, even the Milky Way is dwarfed by IC1101. This supergiant elliptical galaxy lies 1.07 billion light years away in the constellation of Serpens and is a member of the Abel 2029 galaxy cluster. Measuring 5.5 million light years across, it is the largest known galaxy in the Universe and is fifty-five times the diameter of the Milky Way.

Whether it is the sizes or distances in the Universe being considered, it is incredibly difficult to demonstrate either of these on a single scale as the range from smallest to largest is so immense. From the size of the atom to the monstrous elliptical galaxies, or from the relative proximity of the Moon to the far edge of the observable Universe, the extremes are truly astronomical. It is unfortunately a limitation of the human brain that we cannot comprehend such gigantic numbers, but while they may well be beyond the reach of our imagination, the quest to understand them will continue.

May: Northern Hemisphere Sky

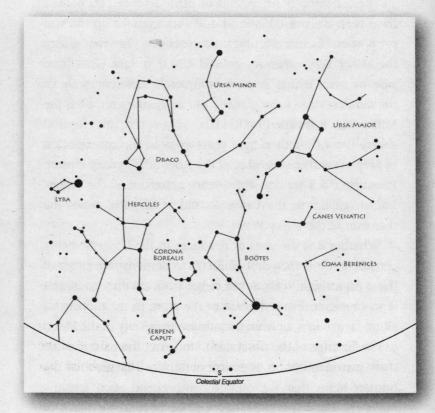

Arcturus is unmistakable in the southern part of the northern hemisphere sky during May. It is by far the brightest star in the region and with its orange glow is easy to identify. Shining at magnitude 0.15 it is a great place to start this month's guide to the night sky. Stretching to the north and slightly to the east is the rest of the constellation of Boötes, home to Arcturus and several other moderately bright stars. Pick out the second-brightest star in the constellation, Izar, which is a beautiful binary star to the north-west of Arcturus and almost 200 light years more distant. The brighter of the two stars can be seen by the naked eye and is an orange giant star with a surface temperature of 4500 degrees. A telescope with an aperture of at least 75mm is needed to detect the fainter companion star, which is a white star shining at magnitude 5.12. Through a telescope the system looks stunning and is well worth seeking out. The rest of the stars trace out the shape of a kite to the north-west of Arcturus, which is supposed to represent a herdsman.

Just to the north-west of Arcturus is the familiar sight of Ursa Major and to the south of its tail is the rather less conspicuous constellation of Canes Venatici. The two brightest stars form a line which runs parallel to the tail, and between its easternmost star, Cor Caroli, and Arcturus is one of the real treasures of the northern night sky, the globular star cluster M3.

M3 was discovered by Charles Messier in 1764 and is just visible to the unaided eye from a dark site. Binoculars show a rather more hazy patch but a telescope with a diameter of at least 20cm is needed to reveal individual stars, although in reality there are around half a million stars in the cluster. The galactic centre which is seen in the direction of Sagittarius is closer than M3, which lies at a distance of just under 34,000 light years.

Another great example of a globular cluster can be found fractionally to the north of the celestial equator, using Arcturus as a guide to spot it. Look to the south-east of Arcturus and find Zeta Boötis, which is fainter than Arcturus and about 10 degrees away (a clenched fist at arm's length measures around 10 degrees against the sky). Follow the line these two stars make towards the celestial equator and, at just over twice the same distance again, the faint fuzzy-looking star is the cluster M5. At 165 light years in diameter it is one of the largest of its kind.

Just a couple of degrees to the north of M5 is IC1101, the largest known galaxy in the Universe, although it is so faint that it takes a telescope about 35cm in diameter to pick it up. To the north-east of M5 is the brightest star in the con-stellation of Serpens, Alpha Serpentis, and it is easily recognized as a moderately bright orange star. It is nearing the end of its life and is burning helium in its core, produc-ing carbon and oxygen, a phase that our Sun will go through in a few billion years' time.

Directly to the north of M5 and Alpha Serpentis is a small constellation called Corona Borealis, which looks like a curve of stars, open to the north and with its brightest star, Alphecca, to its south-west corner. This star is one member of a pair of stars known as an eclipsing binary, like Spica in

Virgo, which means they appear along the same line of sight from Earth and seem to keep eclipsing each other.

To the north-west of Corona Borealis is the northern end of Boötes and directly to the north of Nekar, the second-brightest star in the constellation, lies Edasich in Draco. This is one of the largest constellations and extends out to the east, taking it north of Hercules, and to the west by Ursa Major. Edasich, or Iota Draconis, is an orange supergiant star with a surface temperature of around 4500 degrees, about 1000 degrees cooler than the Sun. With a mass just a little greater than the Sun's, Iota Draconis is thought to be a little ahead of the Sun in its evolutionary cycle and gives us a glimpse of the Sun's future. In 2001 a planet was discovered in orbit around this giant star, and by studying its movement an estimate could be made of its mass at around nine times that of Jupiter. It has a highly elliptical orbit and the planet, now called Iota Draconis B, moves from its furthest point 2.1 times the Earth–Sun distance to its nearest point of 0.4 times. This means that it is very unlikely for any smaller rocky Earth-type planets to exist within the system.

A little further to the north is another orange giant star similar to Iota Draconis but much larger. At a distance of just under 130 light years Kochab in Ursa Minor shines with a brightness equivalent to around 500 Suns, yet it is giving off a lot less light and energy than Polaris to its north. Known as the Pole Star, Polaris lies at a point in the sky directly above the Earth's axis of rotation, so all objects in the sky seem to rotate around it over a 24-hour period. It is the brightest star in the constellation of Ursa Minor, the Small or Lesser Bear, with Kochab just slightly fainter. The reality is that Polaris gives off about five times as much light as Kochab, but at a distance of around 430 light years it appears at a similar brightness in the sky. As well as having two fainter companion stars, Polaris is a Cepheid Variable and, as we have seen, it was stars like this that allowed us to map some of the distances in space. Between them, these two stars mark the far ends of Ursa Minor, which looks like a smaller version of the Plough, with Polaris remaining almost motionless at the end of the handle and Kochab representing the end of the pan along with Gamma Ursae Minoris.

May: Southern Hemisphere Sky

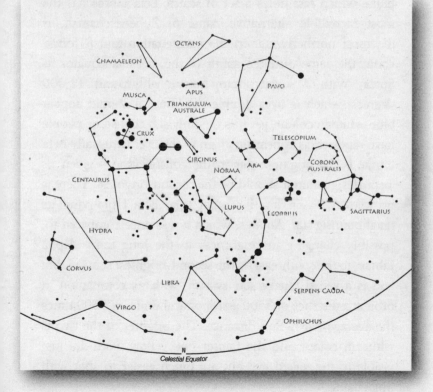

Two prominent stars lie either side of Libra just to the south of the celestial equator in the May sky: Spica in Virgo to the east and Sabik, or Eta Ophiuchi, to the west in Ophiuchus. Spica is by far the brightest star in that region of sky, which is not surprising as it gives off as much light as 2300 of our Suns.

Between Spica and Eta Ophiuchi is the constellation of Libra, which resembles a set of scales. Beta Librae has the most incredible alternative name of Zubeneschamali, is the most northerly star in the constellation and is found about the same distance south of the celestial equator as Spica. With a surface temperature of around 12,000 degrees, which is over double the Sun's, it should appear blue-white in colour, yet it is the only star that some people have reported as appearing green to the eye. Unusually Beta Librae is the brightest star in the constellation – given its 'beta' designation it should be the second-brightest. There is no known reason for this other than Beta Librae and its neighbouring star, Antares, which it was once compared to, possibly changing in brightness in the long term. Alpha Librae, to the south-east, is the second-brightest star in Libra and is a visual binary star system, with its companion in orbit at a distance of 5500 astronomical units, or 5500 times the average Earth–Sun distance. The brighter of the two is white in colour and the fainter one yellow. Both are just visible to the naked eye but are much easier to spot with binoculars.

To the south-west is the constellation of Scorpius and the unmistakable brilliance of the red supergiant star Antares.

This area of sky is well known for its extensive emission and reflection nebulosity, which seems to have been formed by strong stellar winds blowing from Antares and ejecting some of its outer layers into space. The nebulosity is so extensive that it is engulfing nearby Rho Ophiuchi too.

To the south of Libra is the prominent constellation of Lupus, which is said to look like a wild animal being carried by the neighbouring constellation of Centaurus, the Centaur. Ten stars define the shape of Lupus, which is to all intents visible as a rather ragged rectangular shape to the south-east of Antares. Within the borders and just to the north of Zeta Lupi, one of the most southerly stars is a rather lovely double star system called Kappa Lupi, easily observed with binoculars as a pair of blue-white stars with the brightest visible to the naked eye. The pair are not in orbit around each other but they may well have formed out of the same cloud of gas and dust several billion years ago. Scan the sky with binoculars just to the south-east of Zeta Lupi this time and there is a beautiful open cluster of stars, NGC5822, which has around 120 members. This area of sky is rich with background stars of the Milky Way, but look carefully and another fainter cluster, NGC5823, lies just over a degree to the south.

Nearby to the south of Lupus is another unmistakably bright star, Alpha Centauri, which is a Sun-like yellow star, although it has a slightly cooler surface temperature of around 4990 degrees. The star is actually a triple star system composed of two close stars similar to the Sun and a third more distant, a faint red dwarf called Proxima Centauri. This is the nearest star to our own Solar System and is among the smallest stars, only a little larger than the planet Jupiter. Just to the east of Alpha Centauri is Hadar, a blue-coloured star, which gives a clue to its high surface temperature of around 25,000 degrees.

Like its bright companion to the west, Hadar is actually two stars which are individually undetectable to the naked eye. The two have similar properties and orbit each other over a period of around 357 days. They are both giant stars and, as with all stars of this size, have evolved rapidly, taking just 30 million years to get to a stage that will take the Sun around 10 billion years. Quite what will happen to these stars when they finally die is unknown but ultimately their decay is determined by their mass. They are thought each to have a mass around ten times that of the Sun, which could mean they will explode violently as supernovae or will fade gently away, releasing their outer layers into space. Hadar and Alpha Centauri are together known as 'the Pointers', because a line between them and extended through to the east points at Gacrux, or Gamma Crucis, the northernmost star of the Southern Cross. Gacrux and Acrux, the brightest star in the constellation, together point to the South Celestial Pole.

Over to the south-east of Alpha Centauri is a triangular constellation called Triangulum Australe. The angles of the triangle are marked by Atria, the brightest star in the constellation, over to the east, Beta Trianguli Australis at the most northerly angle, and Gamma Trianguli Australis, a star five times the diameter of the Sun, to its south-west. A rather nice yet faint open star cluster is found by following the direction of the triangle as it points to the north. It is given the name of NGC6025 and is easily visible with binoculars about 3 degrees to the north of Beta Trianguli Australis. Another triangle of stars can be seen to the south of Triangulum Australe but it is more elongated. The constellation is called Apus and is thought to represent a Bird of Paradise, but its stars are no brighter than 4th magnitude, making it unremarkable. Stars in the constellation of Octans even further to the south are not much brighter than those in Apus and it is within this constellation that we find the South Celestial Pole. One of its stars, Sigma Octantis, lies within 1.5 degrees of the pole and, although it is only just visible, at magnitude 5.4, it holds the title of the southern Pole Star, or Polaris Australis.

SIX

The Earth, Sun and Moon

O F ALL THE PLANETS, the last one people seem to think
about is our home, the planet Earth. It is unique
among the planets in our Solar System because the conditions are just right for water to exist in its three states, as
a gas, a liquid and a solid, and it is this special property that
makes Earth a haven for life. As it travels around the Sun,
taking just over a year to complete one orbit, it is joined on
its journey by one large companion, the Moon, and between
the three of them they have shaped not only our view of the
night sky but also our very lives. We are fortunate that
the Earth orbits the Sun at an average distance of 150
million kilometres, a distance that makes it not too hot and
not too cold, in fact just right for water, which is essential
for life to thrive.

Like the other inner planets, Mercury, Venus and Mars,
Earth is a small rocky planet and at its centre is a core
separated into a solid inner and a liquid outer layer. It is

thought that convection currents in the liquid outer core are responsible for our magnetic field. The rest of the planet is divided into the mantle and the crust, which are differentiated by chemical composition and geological properties. Above this lies the protective layer of gas we call the atmosphere, which is held in place by the force of gravity. If it was not for this, life on Earth would never have gained a foothold and we would not be here. Compared to the average diameter of the Earth of 12,742km, the atmosphere seems relatively thin at just 100km deep, although this does vary with latitude, being thicker around the equator than at the poles. We know the Earth spins on an axis, making a complete revolution in just under a day (23 hours, 56 minutes and 4 seconds), and it is this rotation that causes the atmosphere to bulge out around the equator. Even the Earth itself bulges at the equator as a result of the planet's rotation so it is shaped like a squashed sphere called an oblate spheroid.

This is not the only 'deformity' the Earth experiences, as there is another, rather more dynamic one that is caused not by the rotation of the Earth but by the orbiting Moon, and we see its effect in the tides. Like all bodies in the Universe, the Moon and Earth exert a gravitational pull on all other objects but the strength decreases with distance. The Earth exerts a pull on the Moon, keeping it in orbit, and the Moon exerts a much lesser pull on the Earth. The pull from the Moon produces a slight bulge on the part of the Earth facing it that, to all intents, stays there while the Earth rotates under the bulge. As a location on Earth passes through this point it experiences a rise in water levels and, to a much lesser extent, a rise in land levels too. Of course, it is the rise in water levels we really notice as the regular high tides we see at most locations. There is another bulge on the other side of the Earth, which is explained by the decreasing strength of gravity with distance. Effectively, the material inside the Earth experiences a greater pull than the surface material on the far side, so it is almost as though the Earth is being pulled away from the surface and oceans, creating the second, slightly lower high tide seen twelve hours later.

By the same process, the pull of gravity on Earth causes tides on the Moon and this rather incredible tidal interaction has two quite surprising effects for our satellite. The bulge on the Moon-facing side of the Earth is not exactly in line with the two bodies, as you might expect. The material the Earth is made from offers some resistance to the bulge lying directly Moon-ward, so as the Earth spins, it drags the bulge a little ahead of the Moon and it is this slight misalignment that has not only trapped the Moon into keeping the same face pointing towards the Earth at all times but is also causing it to slowly move away from the Earth at a rate of 3.8cm per year.

The explanation of these strange phenomena is that the extra bulge exerts a very small but measurable gravitational pull on the Moon. This pull causes the Moon to speed up in its orbit and as a consequence it moves further away into a 'higher' orbit. This effect was experienced by astronauts at first hand while practising docking procedures between two spacecraft in orbit. The spacecraft behind was accelerated to catch up with the one in front, only to find it moved into a higher orbit and would have shot over the top. Slowing it down caused it to drop into a lower orbit, so it took some practice before the astronauts became competent at juggling altitude and acceleration to dock.

There is a very similar explanation for why we only ever see one face of the Moon from here on Earth, although this has not always been the case. In the distant past the Moon would have seemed to be spinning in the sky so we would have seen all of its surface. Millions of years ago, the tidal bulge on the Moon which was caused by the Earth's gravity used to lie slightly ahead of the Earth/Moon line. The pull of the Earth's gravity tugged on this extra lump of material, producing a braking effect on the Moon's rotation. The constant slowing of the Moon's rotation over many millions of years has been such that it now takes the same length of time to rotate once on its axis as it takes to orbit once around the Earth, so we now only ever see one face.

Having just explained why we can only ever see just one side of the Moon, I have to admit it is actually possible to see a little more than 50 per cent of the lunar surface. In fact, 59 per cent of the Moon is visible over a period of time due to an effect called libration, which is a direct result of the way that the Moon orbits around the Earth. Like most orbits, the orbit of the Moon is elliptical. As it travels around an elliptical orbit it sometimes speeds up and sometimes slows down and because of this we can see a little further around the eastern and western horizon of the Moon. Because the orbit is also very slightly inclined to our orbit by about 5 degrees, occasionally the Moon is slightly 'higher' than us and 'lower' at other times so we can see a little further over the north and south poles. The combined effect of this is that the Moon seems to wobble as it orbits us, allowing us to see 59 per cent of its surface, even though it is tidally locked to the Earth.

This elliptical orbit means the distance between the Earth and Moon varies. The approximate distance was first calculated by the Greek astronomer and geographer Hipparchus around the first century BC. He was ingenious enough to use a solar eclipse, in which the Moon passes between the Sun and Earth, to make a pretty good estimate of the distance to the Moon. In 129 BC a solar eclipse was observed as a total eclipse (rather than a partial eclipse) at the Dardanelles, one of the narrow Turkish straits. Yet only four fifths of the Sun was eclipsed (a partial eclipse) at Alexandria in Egypt about 1000km to the south. Thanks to earlier work by Aristarchus, it was already known that the Moon is closer to the Earth than the Sun and so, during a solar eclipse, will pass in front of the Sun and occupy the same point in the sky. Hipparchus understood this and assumed that the peak of the eclipse occurred at the same time at each location. It later turned out that his assumption was incorrect and that the peak occurs at very slightly different times at different locations. Luckily, though, these time variations were too small to have any great effect on his calculations.

Two observations were carried out, one from the Dardanelles and the other from Alexandria. At the Dardanelles, one particular point on the edge of the Moon lay exactly over a point on the edge of the Sun, but at the same time in the view from Alexandria that same point on the Moon did not quite reach the edge of the Sun by an amount equal to about one fifth of the diameter of the Sun. This equates to an angular measurement of about 0.1 degrees (five times smaller than the full moon). We measure the apparent size of things in the sky using this concept of degrees. To understand this, imagine that the sky is a vast sphere surrounding the Earth. Pick a point on the horizon and visualize a line from that point stretching straight up, right over your head and back down to the horizon. Now imagine dividing this distance up into 180 equal parts, and we end up with a measurement equivalent to a degree. Now divide one degree into ten equal portions and one of these portions would equal about one fifth of the solar diameter, a tiny piece of sky, and this was the measurement made during the solar eclipse.

Hipparchus was a great mathematician and developed a particular mathematical technique, called spherical trigonometry, which helped him to deduce that the angle of 0.1 degrees would represent not only the apparent shift in the sky of the Moon based on two different locations, but also the same angular distance between those two points, the Dardanelles and Alexandria. It is likely that Hipparchus did not know the real distance between them at the time but he would have been able to estimate it from knowing their latitude. He did this by measuring the height above the horizon of the North Celestial Pole, which represents the location in the sky to which the Earth's rotational axis points. Currently, Polaris in Ursa Minor marks this spot to within a degree, but the pole changes position around the sky so about 2000 years ago Polaris would have been a little further away. The height above the horizon of the Celestial Pole is the same as the latitude of the observer; for example, the Celestial Pole would lie over your head if you stood at the North Pole, so it would have an altitude of 90 degrees, the same as the latitude from that point. If you stood at the equator it would lie on the horizon and have an altitude of 0 degrees and the latitude from there is 0 degrees. Hipparchus realized this and measurements from his substantial observations of the stars led him to deduce the distance between the Dardanelles and Alexandria was about 9 degrees of latitude.

Now here is the clever bit: Hipparchus understood that the actual distance between the Dardanelles and Alexandria could be calculated knowing the radius of the Earth, which Eratosthenes, as we have seen, had calculated years before. Having worked out the distance between the Dardanelles and Alexandria, Hipparchus could then use all those figures and some fairly simple calculations to establish that the Moon was at a distance equivalent to ninety times the radius of the Earth. There is some historical ambiguity in the reporting of this figure though, as different texts say he calculated the distance to be between 62 and 73 times the radius of the Earth. We now know the distance to the Moon is on average (it varies due to the elliptical shape of the orbit) sixty times the radius of the Earth. Hipparchus had made a number of incorrect assumptions that led to his overestimate of the lunar distance but, regardless of this, to be so close to the actual distance over 2000 years ago is pretty impressive.

The average distance to the Moon is now known to be 384,399km. As it travels around the Earth its appearance seems to change in a very regular way. These are the phases of the Moon, and it is the changing relative angle of the Earth, Sun and Moon that causes them. We see the Moon, and indeed all objects in our Solar System, because they reflect sunlight, so we see a changing amount of the illuminated half of the Moon. For example, when the Moon is opposite to the Sun in the sky we see all of the illuminated or daytime half of the Moon and see it as a 'full moon', but when the Moon is between the Earth and Sun we see none of the illuminated portion and call it a 'new moon'. Between these phases we see varying degrees of the illuminated part, showing us the whole range of phases.

It is interesting that it takes 29.5 days for this cycle of phases to be completed, although the orbit of the Moon takes a little less at 27.3 days. It is not unreasonable to think that the complete cycle of phases would take the same period of time for the Moon to complete one orbit of the Earth, but there is a simple explanation for this discrepancy. It takes 27.3 days for the Moon to complete one orbit of the Earth, but by the time the Moon has got back to its starting point, the Earth itself has moved a little further around its own orbit. If the Earth was stationary in space then it would take 27.3 days for one cycle of phases, but from the moving vantage point of the Earth it takes 29.5 days.

In comparison to the Moon's orbit around the Earth, our own orbit around the Sun takes 365.26 days. We consider this journey around the Sun to mark a calendar year, although we count it as 365 days. With the extra 0.26 of a day things would very slowly get out of step, so every four years, at a leap year, we put an extra day into our calendar during February. An additional adjustment is applied during century years, too, in order to keep things in step.

The calendar year, as we know, is split up into four seasons: spring, summer, autumn and winter. These seasons you might think are the result of the Earth's elliptical orbit around the Sun, so when it is at its closest we experience summer, and winter at the furthest point. This is not the case. The seasons are caused by a phenomenon that is shared among all the planets, a tilted axis of rotation. The tilt is measured in respect of the plane of the Earth's orbit around the Sun and currently it measures 23.4 degrees from vertical. This angle varies over a 26,000-year cycle as the axis of the Earth wobbles much like a child's spinning top. When around June/July each year the northern hemisphere of the Earth is tilted towards the Sun the northern hemisphere enjoys summer, whereas the southern hemisphere is tilted away and experiences winter. During December/January the southern hemisphere is tilted sunwards and enjoys summer, while northern temperatures drop to their winter levels.

With the Earth orbiting the Sun and the Moon orbiting the Earth, there are occasions in the year when the three lie in a perfect line. When this happens, we see an eclipse. If the Earth is in the middle and blocks sunlight from reaching the Moon we see a lunar eclipse but if the Moon is in the middle and blocks sunlight from reaching the Earth we see a solar eclipse, arguably one of the finest and most spectacular naturally occurring events. As we saw earlier, the orbit of the Moon is tilted with respect to the orbit of the Earth around the Sun so we do not see eclipses every full moon and every new moon as the three objects do not line up perfectly.

Eclipses of the Sun provide a great opportunity to study its rarefied outer atmosphere, called the corona. If it were not for the eclipse, the relatively faint outer shell of gas would be completely outshone by the much brighter photosphere. It is necessary to take great precautions when observing the Sun due to the intense levels of radiation coming from it. Looking at the Sun with the eye alone will lead to permanent damage, even blindness. There are safe ways to study the Sun using expensive filters, but the most straightforward method is to project the image through a pair of binoculars or telescope onto a piece of card. Simply placing the card about 30cm away from the eyepiece will produce a nice projected image of the Sun upon which you can see detail on the photosphere. If you are using a telescope with an aperture over 10cm, it is a good idea to cut a 7.5cm circular hole out of a piece of card large enough to cover the end of your telescope so that you reduce the amount of sunlight and heat entering the tube.

The photosphere referred to above is the visible surface of

the vast sphere of gas that makes up the Sun. Like all stars, the Sun was formed out of a huge cloud of primarily hydrogen gas and dust which we call a nebula. Over many billions of years, gravity caused the gas to compress to such an extreme extent that the hydrogen atoms started smashing together and fusing into helium atoms. As a helium atom is formed, a small amount of heat and a small amount of light are given off and it is this moment that marks the birth of a star. Deep in the heart of the Sun, fusion processes like this are happening constantly and the energy generated in the core gives us the heat and light we rely upon every day.

The temperature in the core of the Sun is around 16 million degrees and it is here that the sphere of helium is growing as the hydrogen atoms are fused together. As the fusion process continues, the thermonuclear pressure, as it is called, exerts an outwards pushing force, which for most of the Sun's life is balanced by the pull of gravity trying to collapse it. Eventually, the fusion of helium atoms into carbon and other elements begins in the core and the thermonuclear pressure starts to build even further. When it becomes greater than the force of gravity the Sun will expand into a red giant and in the end die.

For about 10 billion years, the Sun will remain stable and it is in this phase that we can see the Sun today. It measures around 864,000km in diameter and could swallow up over a million Earths. When observed safely through solar projection it is possible to see details in the photosphere and some of the easiest to see are the sunspots. Their appearance seems to follow an eleven-year cycle from minimum to maximum activity and studying them gives a clue to the nature of the Sun's magnetic field.

As the Sun rotates, the gas, which acts more like a thick liquid called a plasma, drags the magnetic field lines around with it, but because the polar region rotates slower than the equatorial regions, the field lines get wound up like a coiled spring. They can only take so much stress and occasionally snap, bursting through the photosphere and temporarily blocking energy and light from below so we see a dark sunspot. Eventually, after eleven years, the field lines get so tightly wound up that they snap back into place, giving rise to the eleven-year sunspot cycle.

If the Sun is studied through specialist filters which significantly restrict the wavelength of visible light, it is possible to see a different, almost violent Sun instead of the rather more serene view that solar projection offers. Where the magnetic field lines producing sunspots have broken through the photosphere, they give rise to spectacular prominences as the field lines drag plasma away from the Sun in dramatic loops. It is not unusual for these prominences to become disconnected from the Sun and send solar material off into the Solar System.

Aside from these prominences, and heat and light, the Sun is also constantly emitting a stream of invisible, electrically charged particles called the solar wind. The origins of this phenomenon are still uncertain but it is thought to come from the corona, the outer atmosphere of the Sun. There are two components to the solar wind, the slow and the fast, which move through the Solar System at

400 and 800km per second respectively. The quantity of charged particles in the solar wind varies along with the eleven-year sunspot cycle, but at peak activity we see bursts of the wind here on Earth.

On its arrival we can be treated to some truly incredible sights as the electrically charged solar wind gets channelled towards the north and south poles by our magnetic field. As it cascades into the upper atmosphere, the electrical properties of the wind give energy to the gas in our atmosphere and cause it to glow just like a neon tube. Usually these aurora displays are restricted to the far north or south, but if there is a significant burst of solar wind – for example, during events known as coronal mass ejections – then lower latitudes can be treated to the show. In the northern hemisphere the aurora are called the northern lights, or aurora borealis, and those seen in the southern hemisphere are called the southern lights, or aurora australis.

Another transient spectacle that can be enjoyed in the night skies of our home planet is given by meteor showers. As the Earth travels around the Sun, it intercepts pieces of interplanetary rock and dust, which can vary in size from millimetres to metres. Before they can hit the surface they must first fall through our atmosphere and as they fall they heat the air ahead of them, making it glow. This produces the characteristic flash of light known more commonly as a shooting star. Generally the smallest pieces will burn up high in the atmosphere and we call these meteors, but those that survive the fiery plunge and land are known as meteorites.

It is possible to see meteors at any time of year. Those that are solitary travellers through the Solar System and appear as random unexpected meteors are called 'sporadic'. More predictable are the meteors associated with any of the twenty or so meteor showers that peak at around the same time each year. These showers are seen when the Earth travels through the orbit of a comet and it is the comets, constantly shedding material, that seed their orbits around the Sun with dust particles known as meteoroids. Every year at roughly the same time, the Earth will plough through this meteoroid field and we see a sudden surge in meteor activity, with the increase dependent on the density of meteoroids along the comets' orbits. As the shower peaks we can see anything from ten to several hundred meteors per hour.

It is easy to tell if meteors are from a specific shower as they all seem to come from one point in the sky, the shower's radiant. In reality the bits of rock are all travelling in parallel paths much like cars on a long straight road but due to the effect of perspective they all seem to come from one same distant point. The location of this point is what gives the shower its name, so, for example, meteors from the Pi Cetids shower, which peaks late in June, all seem to come from a radiant located in the constellation of Cetus.

The only way to get a good view of a meteor shower whose peak may last for an hour or so is through perseverance. Wrap up warm, get settled on a sun lounger or other suitable outside seat, lie back and look up. It is best not to look in the direction of the radiant or in the opposite part of the sky, but somewhere in between should bag you a few meteors. Be aware of the Moon in your quest to witness a meteor shower though, as when it is full its light can blot out all but the brightest stars and meteors.

From your vantage point, warm on a comfortable chair, it is easy to get lost in the stars, but take time to enjoy the Moon without a telescope or binoculars and look at it as our ancestors would have done thousands of years before us; and, with caution, appreciate the Sun as our own life-giving star; not forgetting our home, the planet Earth, which shares a very special place with both.

June: Northern Hemisphere Sky

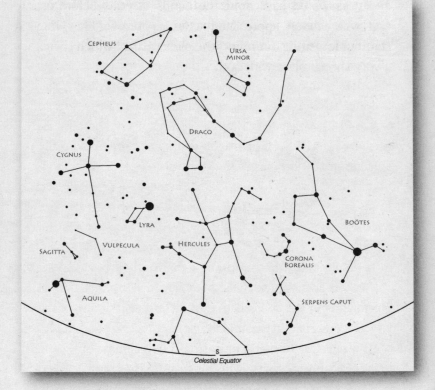

In June, the Sun is at its northernmost point in its journey around the sky, which means the days in the northern hemisphere are at their longest and the nights their shortest. For observers at high latitudes it never really gets dark, so the summer nights are best left for the brighter objects and these provide a great starting point for this guide. Vega in Lyra is in the south-east and Arcturus in Boötes to the south-west. To the north and either side of them both is the brightest star in Ophiuchus, called Rasalhague. Ophiuchus is the thirteenth sign of the Zodiac and its southern stars are found along the ecliptic, the path that the Sun, Moon and planets all broadly follow.

Move to the west and south slightly from Rasalhague to first find the orange star Kappa Ophiuchi, and then a little further on to the slightly brighter Alpha Serpentis. Scan the area around 10 degrees to the south-west and locate a faint, fuzzy star. Through binoculars the fuzzy star appears just a little brighter and larger, but with the larger optics of a telescope, M5 starts to look like a globular cluster. With a 10cm telescope, the brighter stars of the cluster will appear individually rather than forming a part of the fuzz.

Two other globular clusters are visible, to the north of Ophiuchus in the constellation of Hercules. It is one of the largest constellations dominating the June sky and represents a kneeling hunter. For northern hemisphere observers he appears upside down with his head to the south marked by the 3rd magnitude star Rasalgethi. The brightest star in the constellation can be found to the west and slightly north of Rasalgethi and is called Kornephoros, which means 'club bearer' and refers to the weapon Hercules has raised above his head. As with many stars in the sky, it is surprising to find that such relatively faint objects are often giving off more energy than the Sun, which appears much brighter in the sky. This is the case for Kornephoros, as it shines with the equivalent output of 175 Suns and is nearly twenty times as large.

Just to the north of Kornephoros is the second-brightest star in Hercules, Zeta Herculis, and it marks the corner of a pattern of stars known as the Keystone. This forms the body of Hercules and looks like a slightly misshapen square stretching up to the north-east from Zeta Herculis. Along the western edge of the Keystone, north of Zeta Herculis and about two-thirds of the way from Zeta and Eta Herculis, is M13, perhaps the best globular cluster in the northern sky. It is often referred to as the Great Hercules Cluster, although its designation M13 tells us it is the thirteenth object in the Messier Catalogue. It has a magnitude of 5.8 so is just visible from a dark location with the naked eye but a small telescope will transform M13 from a fuzzy blob into a seething mass of glittering stars.

At a distance of around 25,000 light years M13 is one of the closest globular clusters and this is why it is so spectacular in the sky. Any observation of this stunning object will easily show that it contains thousands of stars, although actual estimates range from hundreds of thousands to a million stars, many of which are among the oldest stars in the Universe. It is thought the cluster dates back 12 billion years, which makes it fractionally older than the

Milky Way, although it is home to one young blue star that, it turns out, was gravitationally captured by the cluster when it wandered too close.

Pi Herculis marks the north-west corner of the Keystone with Eta Herculis at the north-east corner. Forming a triangle to the north with these two stars as the base is the other globular cluster in Hercules, M92. It shines at magnitude 6.4, which means it is just beyond the limit of visibility to the naked eye and requires at least binoculars to be revealed. Moving from M13 to M92 nicely demonstrates that larger telescopes will allow fainter and fainter objects to be seen, although M92 is not in fact much fainter than M13. M92 is often overlooked, being overshadowed by M13, but nonetheless it is a beautiful object and well worth hunting down.

To the north of M92 are two of the brightest stars in the constellation of Draco, marking the serpent's head, Eltanin to the east and Rastaban to the west. Both are clearly orange in colour and similar in brightness, although this will not always be the case. Eltanin is heading towards us and in around 1.5 million years will pass within about 30 light years of the Earth, making it one of the brightest stars in the night sky.

Kochab is another orange-coloured star and can be seen to the north-west of Eltanin and Rastaban in the constellation of Ursa Minor, which resembles the Plough in Ursa Major. The tail of Ursa Minor is marked by the 2nd magnitude star Polaris, which is the nearest star to the North Celestial Pole visible to the naked eye. As discussed in

Chapter 3, it is at this position that one of the axes on an equatorially mounted telescope needs to point for it to be 'polar aligned' and therefore capable of following the motion of objects across the sky with a motor.

The axis of the Earth's rotation wobbles in space and much like a slowing spinning top it traces out a circle in the sky over many thousands of years. This means that Polaris has not always been the Pole Star: in about 12,000 BC the Earth's axis was pointing towards Vega in Lyra, which is the bright star to the east of Hercules. This star also marks one corner of perhaps one of the most prominent arrangements of stars in the northern sky, the Summer Triangle. Found to the east of Hercules, it is not a constellation in its own right but simply a shape made from three of the most noticeable stars in the sky during the summer months. Vega marks one of the corners of this giant celestial triangle, with the others marked by Altair in Aquila to the south and Deneb in Cygnus to the east, the Milky Way stretching through it from north-east to south-west. In the centre of the triangle is a star called Albireo, which marks the head of Cygnus, the swan, and is just visible to the naked eye. Without any optical aid it looks like a single orange star, but it has a companion star, blue in colour, which is bright enough to make it easy to spot from a dark location. In reality, the extra magnification of a pair of binoculars or a telescope is needed to allow both stars to be seen properly. Even through small telescopes, the contrast between the two colours is stunning and well worth a look.

June: Southern Hemisphere Sky

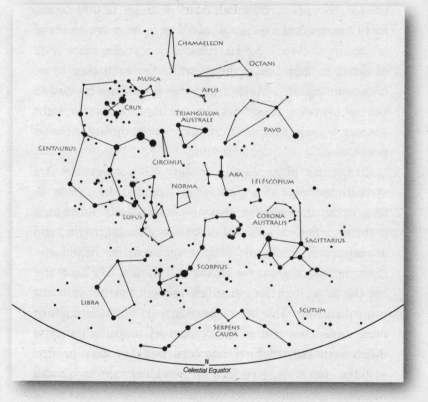

Antares, the brightest star in the constellation Scorpius, sits prominently to the south of the celestial equator in the June sky and glows with an unmistakably red light. Its name means 'Rival of Mars', referencing its striking red colour, but invisible to the naked eye is a fainter companion star known as Antares B, which is a hot blue star with a surface temperature of around 18,000 degrees. Observing Antares B with the naked eye is pretty much impossible due to its proximity to Antares, but a telescope of 15cm or more is capable of separating the two. Larger telescopes reveal a finer level of detail, so increased aperture will allow both stars to be seen more readily, although they can sometimes be viewed through smaller telescopes on the rare occasions when Antares is just obscured by the Moon during a lunar occultation.

To the east of Antares by a mere 1.3 degrees, M4 is a stunning example of a globular cluster, and through some telescopes with a wide field of view both objects can be seen at the same time. It is one of the easiest globulars to find and at magnitude 5.6 should just be visible to the naked eye. Pointing binoculars at the cluster will show it as a fuzzy star but the larger aperture of a telescope will start to show the individual stars. Due to its proximity at 7200 light years, even something as small as a 10cm telescope will start to differentiate the brighter stars – in fact, M4 was the first globular cluster resolved into its individual stars by Charles Messier.

A curve of stars representing the tail of Scorpius extends to the south-west from Antares, terminating in a couple of beautiful stars, Shaula to the west and Lesath to the east. Just to the north-west of Shaula is a different type of stellar cluster, designated M7 but also bearing the name Ptolemy's Cluster, after Claudius Ptolemy, who referred to it as 'the little cloud following the stinger of Scorpius'. The cluster is found on the western edge of the Milky Way and to the naked eye appears as a region with a higher concentration of stars. The larger aperture of a telescope reveals around eighty-five stars, with the brighter stars near the centre. There are many more fine examples of globular clusters, particularly around Ophiuchus to the north of Antares.

Halfway along and to the west of a line between Antares and the brightest star in the sky at this time, Alpha Centauri, lies a rather faint and inconspicuous constellation called Norma. It is named after a scientific instrument whose full name was Norma et Tegula, meaning a level or square. The small constellation is in the shape of a kite pointing to the north and the stars at its corners are all around 4th magnitude. Epsilon Normae marks the most westerly corner and is a nice binary star system whose components are separated by around 400 light years. The brightest of the stars is magnitude 4.5, which makes it visible to the naked eye, but its companion, at magnitude 7.5, requires binoculars or a telescope. Each of the two stars is also a spectroscopic binary, which means their companions are only detectable through the study of the stars' spectra.

To the south-west of Norma is a constellation named Ara, which looks like two parallel lines of stars pointing to the south. The brightest star is found on its western edge, and although it has the designation Beta Arae it is fractionally brighter than Alpha Arae, which is just to the north. Beta Arae is classed as a supergiant star with a surface temperature of around 4580 degrees. With a diameter almost forty-six times that of our Sun, if they swapped places Beta Arae would swallow up Mercury and stretch halfway to the Earth. Located between Ara and Norma is what seems to be a region of the Milky Way with a slightly higher density of stars. In reality, it is a galactic or open star cluster known as NGC6193 sitting within the plane of our galaxy. The cluster is visible on clear dark nights and covers an area of sky almost twice that of the full moon and is thought to have formed just 3 million years ago, surrounded by gas which is visible as the emission nebula NGC6188. This is young in comparison to many other objects in the sky, such as the globular cluster NGC6352 to the north of Alpha Arae, believed to be 2 billion years old. NGC6352 is too faint to be seen with the naked eye but scanning the sky with binoculars about a degree from Alpha will reveal a small fuzzy star, though the larger aperture of a telescope is needed to show the individual stars.

Another globular cluster, NGC6397, is found to the south of Alpha Arae and is just visible to the naked eye. It lies around 7200 light years away and is the second-nearest globular cluster, beaten only by M4 in Scorpius to the north, which is 400 light years nearer. The cluster is thought to be home to almost half a million stars, which are densely packed within the core. The June southern sky contains many globular clusters, which are all distributed around the

Milky Way in a vast halo. This is easily seen as the majority are found along the borders of the Milky Way as it arches across the sky.

Due south-east from Ara is the bright yellow-white star Proxima Centauri and to its east another star, which is not so bright, Hadar. Forming an equilateral triangle with these two stars to the south is a 10th magnitude galaxy called the Circinus Dwarf Galaxy. Unlike most galaxies this one is visible in the direction of the Milky Way, which usually obscures the light from distant objects. Regardless of its moderate brightness it can be easily viewed with small telescopes, but an aperture of at least 20cm is needed to see it in detail.

To the west and a little to the south of Alpha Centauri is the small constellation of Triangulum Australe, which as its name suggests is triangular in shape. Its brightest star is found at the western point of the triangle and is a giant orange star called Atria. Above it is the smaller and fainter constellation of Apus, which represents a bird. It looks like a slim, elongated triangle marked by the stars Beta, Gamma and Delta Apodis, with Alpha Apodis extending out to the east. These two groups of stars lead away from the Milky Way, which can be seen running from the south-west to the north-east, and from a dark location away from artificial lighting it looks stunning.

SEVEN

Transient Stuff

MANY PEOPLE HAVE the impression that the night sky is static and unchanging, but aside from the rather more subtly changing views of the planets and stars there are a number of other more obvious things to look out for. Some of the events covered in this chapter are easy to predict, while others give very little warning and rely on a little bit of luck if you are to witness them, so they range from the clockwork precision of eclipses to the unpredictability of noctilucent clouds. The common factor in all of these phenomena is the Earth, either its position in space or how it interacts with its environment.

Among the easiest to study, and moderately easy to predict, are the yearly meteor showers that grace our skies. There are about twenty reliable annual showers and they generally originate from comets. These 'dirty snowballs' have solid cores made from ice, dust and rock and, for most of their lives, lurk in the far reaches of the Solar System. Here, it is cold and dark and they only venture to the inner

Solar System when disturbed from their remote orbit. For many years it was thought the Kuiper Belt was the origin of the 'periodic' comets that regularly visit the inner Solar System, but recent studies show the Belt to be fairly stable and home to most of the outer minor planets instead. We now believe comet nuclei to come from a region called the 'Scattered Disc'. The members of the Disc all seem to have fairly unstable orbits, most of which are highly elliptical, with a closest approach to the Sun of around thirty times the Earth–Sun distance (the astronomical unit) and at their most distant as far away as a hundred astronomical units.

Another region of the Solar System thought to be home to the long-period comets, as they are known, is the Oort Cloud, which is believed to be a spherical cloud of icy bodies about 50,000 astronomical units from the Sun. At this distance, it is quite conceivable that a passing star could dislodge one of the Cloud's icy bodies, sending it to the inner Solar System before it heads back out into the depths, maybe to return many thousands of years later or maybe never again.

Regardless of whether a comet is the short-period or long-period variety, occurring either more or less frequently than every 200 years, the way it behaves once in the inner Solar System is broadly the same. The increase in heat as it journeys towards the Sun causes the ice to evaporate straight into a gas, a process known as sublimation. This leads to the formation of a vast gaseous halo, or coma, which surrounds the solid central nucleus, and as the ice sublimates it dislodges pieces of dust and rock, scattering them along the

NGC253 is a stunning spiral galaxy in the constellation of Sculptor. It is also known as a starburst galaxy which means it is going through an intense period of star formation. (*Image taken by The Kingsley School*)

NGC2442 is a spiral galaxy 50 million light years away in the constellation of Volans, discovered by Sir John Herschel in 1834. (*Image taken by Mark Thompson*)

M22 was one of the first globular clusters to be discovered in 1665 and lies just under 11,000 light years away from us. (*Image taken by Moreton Hall School*)

M33 is a spiral galaxy that lies about 3 million light years away from us. Some keen-eyed observers report being able to see it with the naked eye, making it one of the most distant objects visible without a telescope. (*Image taken by Shoeburyness High School*)

M83 is known by its other name, the Southern Pinwheel Galaxy. It is one of the closest and brightest barred spiral galaxies to us, making it ideal for amateur observation. (*Image taken by Berlin International School*)

M93 is a beautiful open cluster in the southern constellation Puppis and is thought to be around 100 million years old. (*Image taken by Shoeburyness High School*)

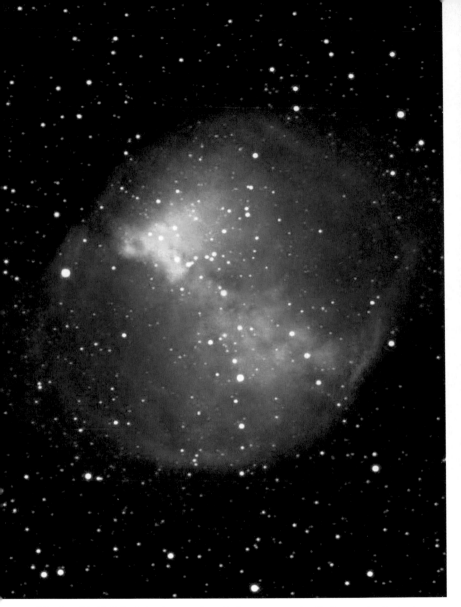

The Dumbbell Nebula is the remains of a lower mass star that came to the end of its life. This was the first and finest example of a planetary nebula discovered. (*Image taken by Loughborough Grammar School*)

The Eagle Nebula is visible to us because of the hot young stars it has produced. As energy pours out of them they sculpt the shape of the nebula, which is said to resemble an eagle.

(*Image taken by Czacki High School*)

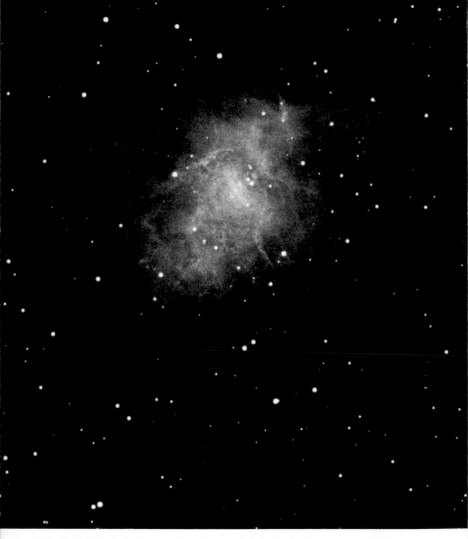

The Crab Nebula is the remains of an exploded star that was seen in 1054 by Japanese, Chinese and Arabian astronomers.
(*Image taken by the Hereford Cathedral School*)

The Omega Nebula is also known as the Swan Nebula because some of the glowing gas clouds resemble a swan on the river. It was discovered by Philippe Loys de Chéseaux in 1745. (*Image taken by Newchurch Community Primary School*)

Omega Centauri is the finest example of a globular cluster in the northern and southern skies. Visible to the naked eye, the cluster is one of the few of its kind that can be seen without optical aid. (*Image taken by Queen Mary College*)

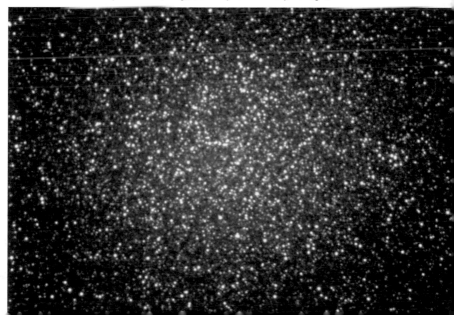

Centaurus A was formed when two galaxies collided leaving the new galaxy in turmoil. At 11 million light years away, it is the closest active galaxy to us. (*Image taken by Portsmouth Grammar School*)

NGC891 is a beautiful example of an edge on spiral galaxy found in the constellation Andromeda. The dust in the galactic disk can be clearly seen. (*Image taken by Kenton Primary School*)

orbit. The solar wind, which is essentially a stream of electrically charged particles from the Sun, pushes against the coma of the comet forcing it 'downstream' and forming the comet's trademark tail. We do know the orbital characteristics of a great number of comets but, even so, they can still offer some surprises. Their visual appearance in the sky is very hard to predict and sometimes they can just end up being very disappointing. On occasions something quite special happens and we are treated to events like the amazing impact on Jupiter of Comet Shoemaker-Levy 9 and the stunning apparition of Comet Hale-Bopp in 1997.

As we have seen, when the Earth passes through the orbit of a comet it sweeps up the cometary debris and this gives rise to the phenomenon of the meteor shower. On any night of the year it is possible to see random one-off meteors that are not related to any shower but are just isolated pieces of interplanetary rock or dust which happen upon the Earth by chance. These are called sporadic meteors and there is no way to predict them. This is quite the opposite of the meteor showers that light up our skies throughout the year, and it is because we have a pretty good understanding of the comets which cross our orbit that we know when they will happen each year. It is true that we cannot predict with much certainty how spectacular, or not, the showers will be since this is related to the distribution and density of dust along the orbit. The only thing we can say with a little confidence is that the comet itself passing by, just before the Earth intercepts its orbit, is likely to give rise to a pretty decent display.

Visually, meteors forming a display will all seem to appear from one point in the sky, known as the shower's 'radiant'. The constellation the radiant falls within gives the shower its name; for example, the radiant of the Leonids is in Leo, that of the Cygnids is within Cygnus and the radiant of the Perseids lies in Perseus. The likely level of meteor activity seen during a shower is explained in the term 'zenithal hourly rate', or ZHR, which simply means that if the radiant were directly overhead at the zenith, then the specified number is the estimated number of meteors seen per hour from a dark moonless sky. This can range from a mere handful to several hundred per hour, but remember that this is under perfect conditions – the radiant is rarely overhead so some will be lost below the horizon, and moonlight often interferes with your view, as does artificial lighting from your observation point.

To see a meteor shower at its best you generally need to be out after midnight, since this puts you on the forward-facing side of the Earth as it hurtles through space and through the meteor debris. It is a little like driving through a swarm of flies: you'll see a lot stuck to your car windscreen but your rear window will be clear. The real optimum conditions are for a shower to peak after midnight, for the radiant to be overhead and for there to be no moonlight or light pollution – then you may be in for a real treat.

During either meteor showers or the odd sporadic meteor you may get to see a fireball, which is essentially just a very bright meteor. Fireballs can be so bright that they cast shadows and momentarily illuminate the landscape. The chances are that these objects are large enough to actually land, and will therefore be termed meteorites. There are many factors that determine whether they will land or not: rock size, speed of travel, angle of path through the atmosphere, composition and many others. A common mistake is to believe that those that do land will be hot and sit smouldering away, but in fact they are more likely to be cold to the touch. As they fall through the atmosphere at speeds in excess of 50km per second, they crash into atoms of gas. This 'impact' heats up and dislodges material from the meteor, in a process known as ablation; this material and the atoms of gas from the atmosphere are broken up into charged particles that give off visible light. Generally this is restricted to the area around the meteor as it falls but occasionally it is possible to see a trail behind the meteor. As the meteor punches through the lower atmosphere the gas in front of it is compressed and heats up, and on rare occasions this compression gives rise to a shockwave that can be heard as a sonic boom.

One of the most interesting annual displays is the Leonid shower, which usually peaks each year around 17 November. Often the shower is quite mediocre with just a few meteors at its peak, but every thirty-three years or so there is a dramatic increase in the number of meteors and in excess of a thousand are seen, revealing to us the beauty of a meteor storm. The particles that cause the shower come from the comet Tempel–Tuttle, which orbits the Sun once every thirty-three years; when the comet crosses the Earth's orbit there will be a higher than normal amount of debris, leading to storm-level displays. The great thing is that you do not need any expensive equipment to enjoy a meteor shower; whether you are observing the Leonids or another shower the technique is the same: wrap up warm, get comfortable, lie back, look up (away from the radiant), relax and wait.

Perhaps even easier to observe than meteor showers are the eclipses of the Moon. They occur when the Earth lies directly between the Sun and Moon, blocking sunlight from reaching the Moon. In reality a little light does illuminate the Moon when refracted as it passes through the stratosphere. Scattering effects cause the red light to make it through, where it lights up the Moon with a ghostly red glow during the total phase. If the upper atmosphere is full of dust, e.g. from volcanic eruptions, then the eclipse will be dark, but if the atmosphere is relatively clear then a brighter eclipse will be seen.

During the eclipse, the Moon effectively passes through the shadow of the Earth, which can be divided into two parts: the darker central umbra, in which the Earth blocks all light from the Sun, and the penumbra, where some sunlight still penetrates. Three different types of eclipse can be seen, depending on where the Moon falls within the shadow. If it passes through the penumbra, the Moon will darken only very slightly, but the more spectacular eclipses are seen when part or all of the Moon falls inside the umbra. If a partial umbral eclipse is seen, only part of the Moon passes through the umbral shadow and only a portion of the lunar disc turns very dark. If the entire Moon passes through the umbra, the whole Moon would ordinarily disappear from view, but, as we have seen, some light is redirected through the Earth's atmosphere, allowing red light to gently illuminate it. These effects can be stunning, and with the Moon turning a deep blood red it is not hard to see why ancient civilizations saw them as a portent of doom and destruction.

Lunar eclipses can easily be studied and appreciated with just the naked eye and due to their nature they can be seen anywhere on the Earth where the Moon is visible. This may seem to be stating the obvious, but the same is not the case with a solar eclipse, in which the Moon passes between the Earth and Sun. During these incredible displays you must be in very specific locations in order to see the eclipse; the Sun may be visible where you are but this does not mean you will see an eclipse. This is what makes a solar eclipse much more of a challenge to witness than its lunar counterpart and why people will travel thousands of kilometres to see one.

Those who do make the effort will be treated to one of the most amazing sights the Universe has to offer. The reason why solar eclipses in this age are so spectacular is down to chance and it will not be possible to enjoy them in their true splendour for ever. The Sun is about 400 times larger than the Moon and, currently, about 400 times further away from us. This is not just a mathematical curiosity; it means that the Moon and Sun can appear to be exactly the same size in the sky. Because of tidal effects the Moon is moving away from us at a rate of 3.8cm per year, which means that in the far distant past it would have appeared larger in the sky than the Sun and in the future will appear smaller. Eclipses would have been much less impressive millions of years ago than they are now.

Solar eclipses do not always look the same today anyway as the orbit of the Moon is elliptical, which means its distance from us varies so its apparent size in the sky changes. If a solar eclipse occurs when the Moon is at its most distant from the Earth then it will be smaller than the Sun and will not block the whole solar disc from view. At this point we will see an annular eclipse, in which the silhouette of the Moon is surrounded by a ring of light from the Sun. In a few million years' time this will be the only type of eclipse we'll get to see, but at the moment we are lucky enough to be able to enjoy total solar eclipses.

As we saw in Chapter 6, total eclipses are so much more impressive because the Moon is just large enough to block out the light of the Sun's bright photosphere and reveal the fainter yet stunningly beautiful outer atmosphere, the corona. The material in the corona is affected and moved by the Sun's magnetic field so it follows the magnetic field lines, in a way that resembles that experiment from school days with a bar magnet and iron filings. The path of visibility of the eclipse tracks along the surface of the Earth across a fairly narrow corridor so the moment of totality lasts for only a few minutes from any one location.

A word of caution though: observing solar eclipses needs very careful consideration. The light from the Sun is intense and anyone looking at it directly is risking serious damage to their eyes. It is only at the moment of totality during a solar eclipse that it is safe to look at the Sun without any protection. At the moment just before the onset of totality, there is still a tiny glimpse of the bright photosphere of the Sun, which even then is still dangerous and can cause damage. My message is simple: unless the Sun is at the moment of total eclipse it is harmful to look directly at it.

There are safe methods you can use and they certainly do not include using the tiny filters that are supplied with cheap telescopes that fit over eyepieces. These are dangerous and will crack under the intense solar energy, allowing the full force of the Sun's energy to enter your eye. There is a material called Mylar that looks like thin tin foil and, when fitted over the front end of the telescope tube, cuts out enough of the harmful solar energy to allow safe observation. Alternatively you could spend a lot of cash on specialist solar filters or you could project the image of the Sun through binoculars or telescopes onto a sheet of card. You can find more details of this on page 158.

Eclipses of the Moon only happen when it is at the full moon position and eclipses of the Sun only when the Moon is at new moon, yet we do not always see two eclipses each month. As we have seen, the orbit of the Moon around the Earth is tilted with respect to the orbit of the Earth around the Sun, so on many occasions the three objects do not align perfectly for an eclipse and, in reality, the Moon is either slightly above or below the other two objects.

Both types of eclipse are examples of an effect known as syzygy, which means an alignment of three celestial bodies that are bound together by gravity. It is not just the Sun,

Earth and Moon that can align in this way though, as it is reasonably common for the Earth, a planet and the Moon or even the Sun to align. We call these events occultations when an apparently smaller body is blocked by an apparently larger one, or a transit when the smaller object moves in front of the larger. Occultation often refers to events in which the Moon occults distant planets, asteroids or stars, but it is possible for planets to occult stars, although these events are much more rare. The last one of these infrequent events was seen back in 1959, when Venus occulted Regulus in Leo.

Lunar occultations that involve stars are regularly timed by amateur astronomers and they help to fine-tune our knowledge of the Moon's terrain. The lack of a significant atmosphere on the Moon means the light from distant stars is relatively unaffected and almost instantly flicks out when they pass behind the edge, or 'limb', of the Moon. Due to the path the Moon takes around the sky there are four bright stars that it can block: Regulus in Leo, Spica in Virgo, Antares in Scorpius and Aldebaran in Taurus. The most impressive lunar occultation occurs when it passes in front of the Pleiades star cluster in Taurus. When seen through a pair of binoculars, the stars of the cluster disappear one by one before slowly emerging on the other side of the Moon.

Identical in nearly every way are the transits, which are still alignments of three celestial bodies but the larger object is at the back. Transits across the Sun are among the most spectacular, but from Earth only transits of the inner planets, Mercury and Venus, are possible. Other types of transits can be seen when satellites pass in front of their parent planet, such as the Galilean moons which can be seen transiting Jupiter at various times. Transits are not as scientifically valuable as they once were, but transits of Venus across the Sun were once used to estimate the distance to the Sun and timings of the transits of Jupiter's moons were formerly used to calculate the speed of light.

Aside from giving us the backdrop for transits of Mercury and Venus, the Sun is responsible for another phenomenon that comes and goes and peaks with its activity, the aurora. As we saw in the previous chapter, the auroral displays are the result of solar wind, which seems to originate in the Sun's outer atmosphere.

At an altitude of around 80km the electrically charged particles in the solar wind begin changing the energy state of the molecules of gas in the atmosphere, causing them to glow. As the electrons drop back down to their usual energy level, they must get rid of the energy they gained, which is achieved through emitting light. The wavelength and therefore colour of the light released are determined by the gas molecules giving off the light; for example, the green-brown colour is light being released from oxygen molecules and the blue-red light comes from nitrogen gas. The two gases are the major components of the atmosphere and it is their different atomic structures that give rise to the variation in colours.

Unfortunately, many different factors affect whether an auroral display will be seen, including the location on Earth. The North and South Poles generally will see nightly displays of the aurora, but the further towards the equator, the less the chance of seeing them. The quantity of particles in the solar wind is also of paramount importance and the coronal mass ejections, as they are called, which are the most massive outbursts, can increase the chances of seeing the aurora at lower latitudes tremendously. The only advice for catching the aurora displays, other than heading to your nearest polar region, is to check the various aurora alerting services and, when chances of a display seem high, keep an eye on the horizon towards whichever pole is in your hemisphere; but it is important to avoid high levels of light pollution in that direction as they can mask the show.

The eerie noctilucent clouds are another atmospheric phenomenon that can be enjoyed in the darkening twilight sky. Their name means glowing or shining at night and neatly describes their appearance as they seem to hang against the darkening night sky yet are so bright they seem to glow. They can be seen for a period of up to about eighty days, centred broadly on the summer solstice in each hemisphere, and generally only at between 50 and 70 degrees of latitude. The clouds were first observed in 1885, just a couple of years after the eruption of Krakatoa, although it is unlikely that the eruption had anything to do with their formation. Most likely, more people were looking at the sky at this time and simply noticed the unusual cloud display for the first time.

The clouds are composed of crystals of water ice which form high up in the mesosphere, the layer of atmosphere directly above the stratosphere, and are the fractured edge of a polar weather phenomenon called polar mesospheric clouds. The more familiar clouds seen from day to day form lower in the atmosphere in a layer called the troposphere and it is here that tiny particulates form the nucleus upon which water drops condense to produce the clouds. The crystals that produce noctilucent clouds are thought to form around dust particulates from micrometeors or volcanic eruptions, but they also condense directly from water vapour. The curious thing is that there should be very little water vapour in the mesosphere, as it has less than 100 millionths the amount of moisture than in the driest air found on Earth, in the Sahara desert, so their origin is still a mystery. One theory suggests the water vapour may be lifted up from the troposphere below or is possibly the result of chemical reactions, caused perhaps by human activity. The atmosphere is incredibly thin at the altitude of the noctilucent clouds, around 80km, so the ice crystals form at the very low temperature of −120 degrees. The mesosphere is at its coldest of course over the poles and, curiously, during the summer.

The clouds' characteristic glow comes from their high altitude, where they are still being illuminated by sunlight in contrast to the deepening twilight at ground level. Their striking blue-white appearance is the result of absorption of the incoming sunlight by the ozone in the atmosphere. It is hard to accurately predict if and how they will appear, but they are usually to be seen between May and August in the northern hemisphere and November and February in the southern hemisphere. Even though the clouds are part of a polar cloud, they are difficult to see above latitudes of 65 degrees because the sky never darkens enough, and lower than latitudes of 50 degrees the clouds are typically hidden below the horizon.

Far from being unchanging, the night sky is full of new and sometimes unpredictable things to see and, in the case of the phenomena covered in this chapter, nothing more than your own eyes is needed to witness them. Let us not forget, though, that the Universe itself is constantly changing; stars come and go, galaxies collide and evolve, and even planets alter their appearance, so there is always something new to enjoy.

July: Northern Hemisphere Sky

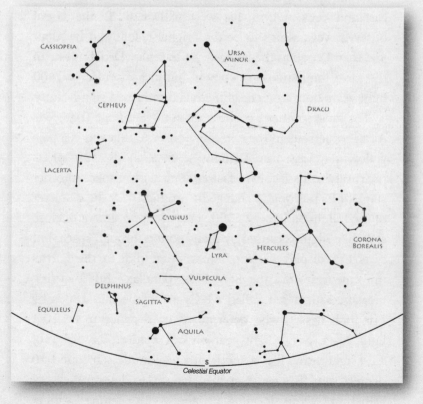

Looking at the sky we can see why our ancestors believed the night sky was perfect and constant, but it is not until you examine it more closely that you realize it is changing. Take the stars in the easily recognized formation to the south-east called the Summer Triangle: they appear to remain the same year after year. The points of the triangle are marked by Altair in Aquila to the south, Deneb in Cygnus further north-east and Vega in Lyra due west of Deneb. To the casual observer Vega seems to be the brightest, followed by Altair and then Deneb as the faintest, yet in reality Deneb is one of the most luminous stars known, but at a distance of 1400 light years from us its brightness is diminished significantly.

The most southerly of the stars in the Summer Triangle is Altair, which sits in the constellation of Aquila and is a pale yellow star just over 16 light years away. Because of its proximity to us it moves fast enough with respect to other stars for its position to change by around twice the diameter of the full moon in only 5000 years. The movement of close stars like Altair is caused by their motion through space, but stars can display an apparent motion due to the Earth's movement around the Sun. This 'parallax' shift was first measured for a star called 61 Cygni in the mid-1800s by Friedrich Bessel, who determined its distance to be 10.4 light years, just 1 light year off our modern-day value of 11.4 light years. It is a subtle 5th magnitude star found 10 degrees (one fist's width at arm's length) to the south-east of Deneb in Cygnus, forming a triangle with Gamma Cygni at the centre of the cross.

Just to the north of Altair is a red giant star over 400 light years away, and north of this is the faint and discreet constellation called Sagitta, which represents an arrow shot by Hercules. The four brightest stars in the constellation are between 3rd and 4th magnitude, making them easy to spot from dark skies. Scan the skies to the north-west of Sagitta and there is a beautiful cluster of stars known as the Coathanger, and as its name suggests the brightest stars form the shape of an upside-down coathanger. Over thirty stars make up this loose open cluster and it is a real treat for binocular observers.

Further north still is another faint constellation, Vulpecula, which is made up of three stars fainter than 4th magnitude forming a shallow triangle with its brightest star, Anser, at the northern point of the triangle. Through binoculars, this star looks like it has a fainter companion, called 8 Vulpeculae, but this is just a line-of-sight effect, with nearly 200 light years separating the two. Through binoculars or the wide field of a finder telescope, the star at the eastern end of the three brightest stars, 13 Vulpeculae, forms a semi-circle with its flat side to the south. Almost halfway along this flat side is one of the real treasures of the summer sky, M27, the Dumbbell Nebula. It is a type of nebula that resembles a planet in small astronomical telescopes, and hence is called a planetary nebula. M27 was the first object of this type to be discovered and is visible with binoculars, but it is stunning through even modest amateur telescopes. Despite their name, planetary nebulae have nothing to do with planets but are giant stars that have reached the ends of their lives and have lost their outer layers of gas to interstellar space. In the case of M27, its famous dumbbell shape is surprisingly common: magnetic field lines, affected by the rotation of the stellar core that was left behind, have sculpted the form we see today. Binoculars show it as a fuzzy star, while small telescopes reveal a little of the structure of the nebula, but the extra power of a large telescope is needed to uncover the remaining core.

Marking the head of Cygnus almost perfectly in the centre of the Summer Triangle and just to the north of the Coathanger and M27 is a star called Albireo, which to the naked eye looks like a single yellow star. Telescopes or even a good pair of binoculars reveal the beauty hiding from view, a stunning double star system with a brighter yellow star and a fainter blue neighbour. The contrast between the two has led to some people reporting the companion star as appearing purple in colour. The whole system is about 380 light years away but there is some doubt as to whether the two are actually orbiting around each other or just happen to appear close at the moment. If they are gravitationally bound it must take in the order of 80,000 years for the completion of one orbit.

Moving off to the north-west of Albireo we get back to the bright and unmistakable star Vega, the brightest star in the constellation of Lyra and by far the brightest star in that part of the sky. It shines with a distinctly blue-white light, indicating that it is a hot young star with a surface temperature in excess of 10,000 degrees. To the north-east of Vega by just a few degrees is perhaps one of the best-known binary stars in the sky, Epsilon Lyrae. Through binoculars it can be seen to have two components, which orbit around each other, but with the greater ability of a telescope to resolve finer detail the two stars can be seen themselves to be binaries, making this a complex quadruple star system. Even telescopes with an aperture of 10cm can split all four stars, although poor-quality instruments may struggle. There is a fifth component star but this is far too close to one of the others for it to be visible directly.

Moving off to the south-east of Vega is a collection of four fainter stars, in the shape of a small squashed square, or parallelogram, making up the rest of the constellation of Lyra. In between the southernmost stars, Sulafat to the east and Sheliak to the west, is M57, another example of a planetary nebula like M27, although visually they are quite different. Unlike M27, M57 is a much more uniform ring shape, hence its common name of the Ring Nebula. It can just be seen with big binoculars as a faint fuzzy star and telescopes as small as 7.5cm will start to reveal the ring structure.

Extending the line between the two southern stars of Lyra's parallelogram to the east leads towards Albireo in Cygnus, and directly between Sulafat and Albireo is one of the fainter objects in the Messier catalogue, M56, an 8th magnitude globular cluster like those studied by Harlow Shapley to determine the shape and size of our galaxy.

Another example of a planetary nebula can be seen to the north-west of Deneb and just to the east of Iota Cygni, which marks the western wing of the swan. Much fainter than M27 or M57, NGC6826 shines at magnitude 8.89, making it a challenge to view with binoculars. Its common name is the Blinking Nebula, which refers to the way it pops in and out of sight if viewers shift their gaze to and from the central star and slightly to one side.

July: Southern Hemisphere Sky

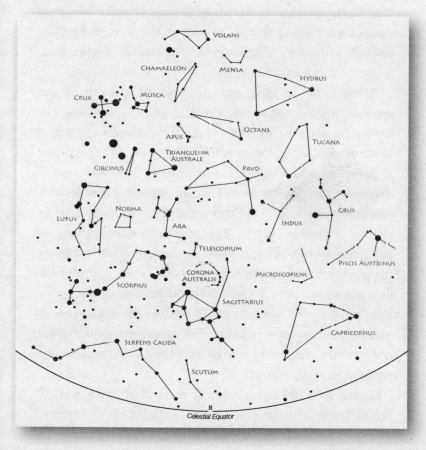

The Milky Way is a prominent feature of the July southern sky, cutting across the celestial equator and heading south. Running across the Milky Way about 20 degrees to the south of the celestial equator is another imaginary line called the ecliptic, which, as I've mentioned, is the apparent path that all the planets in the Solar System broadly follow. It intersects the celestial equator over in the west and passes through Aquarius, Capricornus, Sagittarius, Ophiuchus, Scorpius, Libra and Virgo.

Between the ecliptic and the celestial equator is the second-brightest star in the constellation of Ophiuchus, Eta Ophiuchi, with the majority of the constellation extending to the north-east. Lying directly to the north of Eta are two globular clusters, M12 and M10 (nearest Eta), with only 3 degrees separating the two. They are both just visible with binoculars but require a 15cm aperture telescope to start to resolve individual stars. The southern summer sky is peppered with dozens of globular clusters. Take a look at the positions of the clusters covered in this guide to see how they are found almost exclusively either side of the Milky Way, which runs from the north-west to the south-east. It was by plotting their position that astronomers discovered not only the shape and size of the Milky Way but also our position in relation to it.

To the west of Ophiuchus and embedded in the stars of the Milky Way is the small, faint constellation of Scutum. The stars form the shape of a flattened diamond along the band of the Milky Way, and at its northern end there seems to be a particularly bright patch of light, called the Scutum

Star Cloud. To the south-west is a darker patch of sky, known as the Great Rift. These regions are sculpted by dark interstellar dust clouds, though there is an absence of dust in the Star Cloud, in contrast to the Rift, where an abundance of it blocks distant starlight.

Sagittarius sits to the south of the Scutum Star Cloud and has a few prominent stars. Heading south from Scutum leads first to Nunki, just off the western border of the Milky Way; it is one of the stars in the famous 'Teapot' shape of stars. This unofficial group of stars makes up a great portion of Sagittarius and from the southern hemisphere it looks like the Teapot is upside down. Three other bright stars make up the rest of the handle to the south and, a little further, the more prominent star Kaus Australis marks the base of the spout. Delta and Gamma Sagittarii lie just to the north of Kaus Australis, mark the top of the spout and, pointing to the east, show the way to the centre of our galaxy tens of thousands of light years away

To the east of Nunki is another of the brighter stars in the constellation, Kaus Borealis, which marks the top of the teapot lid. Taking a line from Nunki through this star and on again for the same distance leads to the Lagoon Nebula, a glowing cloud of interstellar gas divided by a dark dust lane. It is visible as a faint glow to the naked eye but telescopes with low magnification grant a much more impressive view.

The bright star just to the south of Nunki is called Ascella and it marks the base of the Teapot's handle. Scan the skies to the west of Ascella about 10 degrees away to find M55, one of the closest globular clusters to us at a distance of

17,300 light years. It is right on the limit of visibility to the naked eye but binoculars will easily pick up the speckled haze of stars, which is a little smaller than the apparent diameter of the full moon in the sky. Its relatively large apparent size is due to a combination of its proximity to us and its actual dimensions, at 100 light years across.

To the south-east of Sagittarius are the bright stars that form the hook-shaped tail of Scorpius. Shaula is the second-brightest star in the constellation and represents the end of the tail, with Lesath just off to its south-east marking the sting at the tip of the tail. Almost at the other end of the constellation is the bright orange-red star Antares, which is one of four bright stars that can be occulted by the Moon. Follow the line of three stars from Shaula to the north-west for about 5 degrees to the globular cluster NGC6541. This globular cluster is on the very limit of visibility to the naked eye from dark locations, at 22,800 light years away. At that distance and direction it is believed to lie just 7000 light years from the centre of the galaxy.

NGC6541 actually lies in one of the neighbouring constellations, Corona Australis. It appears as a semi-circle of faint stars with the open side facing the tail of Scorpius. Starting on the northern side of the curve at the star nearest Scorpius, the fifth star along is Gamma Coronae Australis, which is a great target for smaller telescopes. To the naked eye it appears as a single star, but even small telescopes will reveal two yellow stars which lie 69 light years away. The next star around the curve is the brightest star in the constellation, and it is thought to be twice the mass of the

Sun and to be surrounded by a cool disc of dust that may be evolving into a new planetary system.

Continuing on the line from the stars in the tail of Scorpius leads in the direction of a small, unimpressive and really rather uninteresting constellation called Telescopium, which has a mere two stars to depict the telescope after which it is named. Following the line of these two stars southwards takes us to one of the larger southern hemisphere constellations, Pavo. Its brightest star, Alpha Pavonis, lies to the north-east of the constellation on the border with the faint constellation of Indus, making it fairly prominent in that part of the sky. About 10 degrees to the east of Alpha Pavonis is an impressive example of a globular cluster, NGC6752, which lies 14,000 light years away and is thought to be one of the oldest objects in the Universe at 11.8 billion years. Binoculars will reveal it as a small fuzzy star but even a small telescope will show this cluster well.

Just 5 degrees south of NGC6752 lies a 9th magnitude spiral galaxy named NGC6744, thought to resemble the Milky Way in structure. It is one of the few galaxies visible in the general direction of the Milky Way, which usually obscures galaxies along its plane. Binoculars will struggle to detect the galaxy at all so a telescope will be needed; anything larger than 20cm should be capable of detecting the spiral structure.

EIGHT

Planets Near and Far

LOOK CLOSELY AT THE sky on any night and you might notice one or two bodies that do not seem to twinkle like the rest. Observe over a few weeks and you might also notice that they seem to move against the background of stars. It was observations like these that led ancient stargazers to recognize that there was something special about them. Ultimately they were given a name that described their motion against the stars, the planets, which comes from the Greek word *planetes*, via the Latin *planeta*, meaning 'wanderer'. The motions of the planets had been closely observed for centuries but it was the invention of the telescope that allowed us to start probing their secrets.

The first person to look at the planets with a telescope was Galileo, who was first to spot the rings around Saturn and the moons of Jupiter. These two planets are the largest in the Solar System and, aside from Venus, are generally the easiest to find. If you know in which direction to look for any of the brighter planets then you can spot them since, unlike the stars, they do not twinkle. This is because the stars are so distant from us that they appear as a pinpoint of light whereas planets are closer and their disc covers a larger portion of the sky. The incoming light from both stars and planets gets knocked and bounced around by the movement of gas in the atmosphere making the source appear to move and flicker by a tiny amount, but this effect is much more noticeable with the thin beam of light from a star.

In order to find out which direction to look in to find the planets in the first place there are a number of options, as described in Chapter 2: you could use a planisphere, any of the hundreds of free pieces of astronomy software available on the internet, astronomy magazines or even applications running on smartphones or tablet computers.

Except for Uranus and Neptune, all of the planets are easily visible to the naked eye at some point. The nearest to the Sun, Mercury, is the hardest to spot because it is never far from the Sun in the sky and offers brief views only just before sunrise or after sunset. If you can spot it low in the twilight sky, an amateur telescope is unlikely to show any great detail because it is so small, yet through larger telescopes it may just be possible to see phases like the Moon's, but that is about it. In 1974, Mariner 10 became the first spacecraft to fly past the smallest planet in the Solar System, having first visited Venus. It revealed a world pock-marked with craters and large plains along with evidence of surface movement caused by the shrinking of the planet as the core cools. The core is large compared to the size of the planet, which has an average diameter of 4878km, and due to its iron-rich composition it generates a weak magnetic field like the Earth's. Because of the proximity of Mercury to the Sun, the intense solar radiation has long since removed any atmosphere that Mercury may have once had.

You would expect Mercury to be the hottest planet in the Solar System but, surprisingly, that prize goes to Venus. Planets need an atmosphere to retain heat and, of all those close to the Sun, Venus has the densest. It is not just the density of the atmosphere that drives the high temperatures; the chemical composition has a role to play also. The Venusian atmosphere has high quantities of carbon dioxide and nitrogen and these are the gases that trap the heat from the Sun. The incoming solar energy travels almost un-impeded through the atmosphere before heating up the planet's surface, which in turn re-radiates some of it at a slightly different wavelength. This different wavelength allows the energy to slowly warm up the atmosphere, which is the same process that drives our weather systems here on Earth. The only dissimilarity is that the Earth loses much of its heat back out into space, keeping the temperature moderate, but on Venus that is not possible and instead the atmosphere gets hotter and hotter, giving an average surface temperature of around 460 degrees.

From Earth-based telescopes it is impossible to see the surface of what many consider to be our sister planet, due to the dense sulphuric acid clouds that completely enshroud it. It is left to orbiting spacecraft with radar technology and landers to give us a glimpse of what lies below the cloud, and we find a world scarred by impacts and violent geological activity. It is possible that some of the Venusian volcanoes may have erupted at some point in the recent geological past, evidenced by the presence of sulphur in the atmosphere, although strangely there does seem to be a lack of lava flow on the surface. Studies of the quantity of craters and their structure show, however, that the surface is relatively young at around 300 million years.

As amateur astronomers, what can we hope to see of the second planet from the Sun? Well, of all the planets, Venus is probably the easiest to find as it shines brightly in the twilight just before sunrise in the east or low down in the west after sunset and it is because of this that Venus has earned the name of the morning or evening star. Its bright appearance and moderately low altitude have been the frequent cause of alleged UFO sightings. With the naked eye it is possible to detect a change in the planet's brightness as it moves around the Sun, but with a telescope its phases come into view. The changing relative positions of Venus and the Earth mean that the extent of the illuminated part of the planet visible to us also varies. It is possible as well to detect subtle details in the atmosphere as clouds of sulphuric acid form and dissipate.

The next planet from the Sun is our own home, the planet Earth, which takes a year to complete one orbit. Beyond the Earth is the red planet, Mars, which has some real treats in store for the amateur astronomer. Finding Mars is relatively simple in comparison with some of the other planets because of its striking red colour. Look up its location and, once you know roughly where to look, hunt for a red star. Fortunately there are not many red stars in the sky so Mars will usually stand out. The characteristic colour comes from the presence of a material called iron oxide, which is more commonly known as rust. This red colour is easily visible to the naked eye but through a telescope it takes on a more pinky-brown hue. When the conditions in the Earth's atmosphere are very stable, you can see an amazing amount of detail if you have a modest-sized telescope. Among the most striking and readily observed features on the Martian surface are the polar caps, and they can be glimpsed even through smaller telescopes. Their size varies with the Martian seasons, as they grow in the hemisphere experiencing winter and shrink, even disappear entirely, during the summer. Studies by visiting spacecraft have revealed that they look the same as our own polar caps but they differ in chemical composition, being made almost entirely from carbon dioxide with a smaller quantity of water ice mixed in.

Not so easy to observe is the area called Syrtis Major, which represents the darker subsurface rock that pokes up out of the surrounding dusty plains as a very shallow shield volcano extending 1500km by 1000km. As Mars rotates on its axis, taking about forty minutes longer for one revolution than the Earth does, it is possible to see much of the planet's surface over a period of just a few weeks, so if you cannot spot Syrtis Major on your first attempt, keep trying. It is worth using filters that fit to your eyepiece to sharpen some of the detail; for example, orange or red filters will enhance darker details like Syrtis Major whereas a green or blue filter will enhance the polar caps.

Mars has two Moons, Phobos and Deimos, neither of which is visible through amateur telescopes, but both appear as faint dots of light in larger professional instruments. Their existence was predicted by Johannes Kepler in the first half of the seventeenth century, when he reasoned that if the Earth had one moon and Jupiter had four, then Mars, which sits between the two, must have two moons. It turns out that he was right although his reasoning was wrong, and we now know that Jupiter has many more moons than the four discovered by Galileo.

Moving beyond Mars brings a transition from the rocky, relatively small planets of the inner Solar System to the giant outer gas planets. Jupiter is the next we encounter as it orbits the Sun at an average distance of 778 million kilometres, light from the Sun taking 43.3 minutes to reach it. It is the largest of all the planets in the Solar System – you could fit all the other planets inside and still have room to spare.

Galileo was the first to turn a telescope on Jupiter and during his first few observations discovered a family of four moons in orbit around the planet. These moons, known as Io, Europa, Ganymede and Callisto (collectively the Galilean satellites), were just the beginning as since then another fifty moons have been discovered and a potential further fourteen are still to be confirmed.

As I mentioned earlier, it was the study of the Galilean satellites that in 1676 led the Danish astronomer Ole Rømer to calculate the speed of light. Some years earlier, Galileo had been working on a way to tell the time by the stars that would enable position in longitude to be determined. He came upon a method using eclipses of the Jupiter satellites to tell the time, but making the telescopic observations at sea proved impractical. After joining the observatory of Uraniborg, Rømer studied over a hundred eclipses of the moons, while another astronomer, Giovanni Cassini, was studying them in Paris. Cassini had noticed some errors in the timings in Galileo's tables predicting the eclipses and suggested they may have been caused by light travelling at a finite speed, but it was Rømer who noticed that the times between successive eclipses were getting shorter as Earth approached Jupiter and longer as it moved away. With this information Rømer was able to calculate the speed of light, although for some reason he never published his results. His work led to a figure of around 342,100km per second which is very close to today's figure of 299,792km per second. Not bad for the seventeenth century.

Amateur telescopes reveal not only the four main moons but also surprising details of the planet itself. Jupiter is a gas giant planet, which means it has no solid surface but a rocky core, surrounded by a liquid metallic shell and a gaseous atmosphere. Even with a modest telescope it is possible to see detail in the atmosphere, although visibility varies according to the conditions. The most prominent features are the belts that circle the planet at different latitudes. Strong winds which blow around the planet in opposing directions cause the belts to form and, with gases welling up from below, they change colour to oranges and browns when exposed to the ultra-violet radiation from the Sun. The most easily observed belts are those that circle the planet just to the north and south of the equatorial region.

The opposing winds on Jupiter can cause turbulence, which leads to storms forming. One of the longest-lasting and best-known storms is called the Great Red Spot and it has been raging ever since telescopic observations of the giant planet started. The GRS is the largest known storm in the Solar System and it mimics hurricanes found here on Earth, although it is so large it could swallow up at least two of our home planet!

Jupiter has a lot to offer the amateur astronomer, from the four Galilean satellites to the belts in the atmosphere to the raging storm of the Great Red Spot, but the next planet is the jewel in the crown of the Solar System. Saturn, the sixth planet from the Sun, was the first thing I ever saw through a telescope as a child and the sight that greeted me somehow ignited a fire in my heart that has stayed with me ever since. To see a live view of Saturn's rings, not on television or the internet, nor as a picture in a book, was dazzling.

Saturn is the second-largest planet in the Solar System after Jupiter and it orbits the Sun at an average distance of 1.4 billion kilometres, which is about 9.5 times further from the Sun than the Earth. Like Jupiter, it is visible to the naked eye and is a large gas planet believed to have a dense rocky core, surrounded by a liquid shell of hydrogen and helium which slowly changes to gas with increasing distance from the core. The visible layers of the atmosphere of Saturn are less dramatic than those of Jupiter, with the belts' appearance more subtle and storms occurring a little less frequently.

One of the best-known and most stunning features of Saturn is the incredible ring system. This beautiful sight, visible through even the smallest telescope, is not unique in the Solar System, as Jupiter, Uranus and Neptune all display some form of ring system, although nowhere near as impressive as Saturn's. When they were first observed by Galileo in the seventeenth century he was perplexed as he initially believed Saturn was actually made up of three objects with Saturn the larger, central one. He even referred to the rings as 'ears' in some of his work. Expecting them to be moons accompanying the planet, he thought they never actually moved, unlike Jupiter's, so he was somewhat surprised when a few years later they seemed to shrink and disappear entirely from view, only to return a few months later. What Galileo had in fact observed was the way Saturn's rings change in appearance as the planet seems to wobble in the sky like a giant celestial spinning top as it continues on its 29.5-year orbit of the Sun. Twice during this period we will see a 'ring plane crossing', when the rings of Saturn, which are only a kilometre thick at most, appear edge-on to us and momentarily disappear from our view. Galileo's vanishing moons were a result of this phenomenon.

With today's high-quality telescopes, even bird-watching telescopes, a magnification of just 30x or above will reveal the rings, but a higher magnification will show them in more detail. From the Earth the rings look like a solid structure; higher magnification reveals gaps in the structure, but thanks to space exploration we now know the rings are made up of billions of snowball-sized pieces of rock and ice all orbiting around the planet just as our Moon orbits around us. The gaps in the rings are a clue as to how they maintain their shape and the answer is that larger moons orbit in or near the gaps and their gravitational influence keeps the gaps relatively free of ring debris when compared to the density of material in the rings themselves. One 'gap' is relatively easy to spot through a small telescope and is called the Cassini Division after its discoverer, Giovanni Cassini. It is seen as a dark patch in the rings but studies by the Voyager spacecraft show the gap does have some ring material in it but at a much lower density.

The moons that keep the ring system in place are generally difficult to observe in amateur telescopes but one of Saturn's sixty-two moons, Titan, is fairly easy to spot and can often be seen out to one side of the planet, in comparison looking like a tiny background star. Titan is the largest of Saturn's moons and the only natural satellite in the Solar System to have a dense atmosphere. It is unique, too, in its similarity to the Earth in having what seem to be stable bodies of liquid water on the surface.

Of the two remaining planets, Uranus and Neptune, Uranus is just visible to the naked eye from a dark site and under good conditions, but Neptune remains a target for binoculars or a small telescope. Both planets appear to have a blue hue, which is the result of methane in the upper atmosphere that absorbs the red part of incoming sunlight and reflects the blue. They are made primarily of gas but, unlike Jupiter and Saturn, they have a high concentration of ice in their make-up, leading to them being described as ice giants. This ice is not of the form we would recognize in everyday life, but is actually a hot dense liquid which shares some properties with more conventional ice.

Uranus is at first sight a rather bland world but it hides an incredible secret. Most of the planets in the Solar System spin on an axis that is more or less at right angles to the orbit around the Sun; the Earth's axis, for example, is tilted by just over 23 degrees. Uranus is unique in that its axis is tilted over by 97.7 degrees, which is thought to have been caused by an Earth-sized proto-planet striking it at some point in its early formation. One consequence of this extreme axial tilt is in the level of sunlight and heat the planet receives. At each equinox, Uranus experiences night and day much like any other planet, with the Sun rising and setting during its seventeen-hour day, but during the solstices each pole receives forty-two hours of constant light or dark, depending on which pole is facing the Sun, with the equatorial regions experiencing the Sun low on the horizon. For some reason, the equatorial regions on Uranus are warmer than the poles but we still do not fully understand why.

Neptune, the last of the major planets in the Solar System, is less intriguing than Uranus and many of the other planets, although observations by spacecraft and more recently the Hubble Space Telescope have revealed an atmosphere with interesting features. Of particular interest is the Great Dark Spot, which is similar to the Great Red Spot on Jupiter. It was first observed by Voyager 2 back in 1989 and is a huge storm system rotating around an area of high pressure. The name is often used to refer to a series of short-lived storms on Neptune rather than one particular occurrence, but unlike Jupiter's they seem to form and dissipate over a few years rather than hundreds. There are even thin wispy white clouds of ice crystals high in the Neptunian atmosphere that are similar in nature to the cirrus clouds seen here on Earth.

The unveiling of Neptune was unique among the discoveries of the planets as it was the first to be mathematically predicted prior to being found. Careful studies of the movement of Uranus revealed that something was causing it to be tugged very slightly away from where it should have been. The French mathematician Urbain Le Verrier calculated that this could have been caused by the existence of another planet, gravitationally pulling on it, and even predicted its possible position. But it was the German astronomer Johann Galle who first found and observed Neptune from Le Verrier's predictions in September 1846.

Subsequent observations of Neptune revealed that it too seemed to be affected by disturbance from another planet tugging on it and, following orbital calculations, the search was on again, for a ninth planet. The American astronomer Clyde Tombaugh, working at an observatory in Flagstaff, Arizona, finally discovered Pluto in 1930, although further studies revealed that it did not have enough mass to cause the changes in Neptune's orbit. The search for a tenth planet in the Solar System continued until data from Voyager 2 revealed that Neptune was not in fact being acted upon by another, unknown body.

Until 2006, Pluto was still considered to be a major planet in our Solar System but in that year it was demoted to the ranks of a minor planet. The key reason for this was the continuous discovery of outer Solar System minor bodies, some even larger than Pluto. These discoveries kept growing our family of planets and the lack of an official definition for a 'planet' meant this could continue unchecked. The International Astronomical Union decided to resolve this uncertainty and in 2006 came up with a definition. For an object to be classed as a planet, it had to adhere to three criteria:

1. It must be in an orbit around the Sun.

2. It must have sufficient mass to assume an almost spherical shape.

3. It must have 'cleared the neighbourhood' around its orbit.

Pluto complies with points 1 and 2 but because it has a moon, Charon, which is of comparable size, it is not considered to have cleared the neighbourhood around its orbit. We are now left with a Solar System comprising eight major planets and a whole host of minor ones.

Until 1988, the planets in our Solar System were the only known planets in the entire Universe, but some ground-breaking work by Canadian astronomers led by Bruce Campbell confirmed the existence of a planet orbiting around the binary star Gamma Cephei, 45 light years away.

There was some controversy around its existence but later improvements in detection techniques confirmed it. Discoveries came thick and fast after this, leading to an impressive catalogue of over 700 confirmed exoplanets, as they are known.

Quite a few different techniques are used for detecting exoplanets but the most widely adopted entails measuring the dip of light from a star as an orbiting planet eclipses it by a tiny amount. Another technique is to measure the small shift in a star's position from the tug of an orbiting planet. Of all the discoveries, some of the most exciting are the identification of a planet called Bellerophon, around the star 51 Pegasi, which is just visible to the naked eye, and a system of four planets around Gliese 51B, a star in the constellation of Libra; 51 Pegasi is classed as a main sequence star (which means it is at its most stable) similar to the Sun. Although Bellerophon is a gas giant planet it does at least show that planetary formation around stars like our Sun is common. More tantalizing is the planetary system around the star in Libra (Gliese 581B); of the four planets in orbit, 581C orbits at the correct distance to suggest conditions could be just right for life to evolve. A radio signal has even been sent to the system in the hope of contacting any possible civilizations, although it will not arrive until 2029.

Of all the hundreds of exoplanets around other stars, only about forty have actually been imaged directly. Most of the others are hidden behind the glare of the parent star. Unfortunately, it is going to be some years before we fully understand the conditions on these alien worlds, but what these discoveries have shown is that planet formation is in no way unique to our Solar System; instead it seems to be commonplace around the galaxy and perhaps even the Universe.

August: Northern Hemisphere Sky

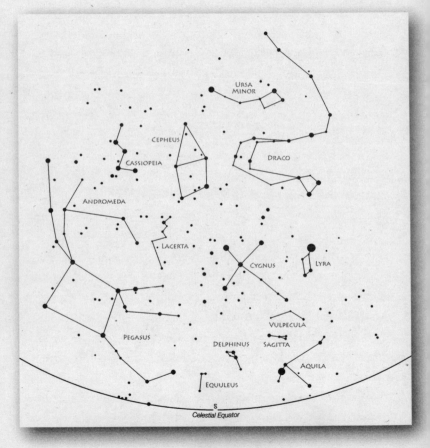

Two faint constellations, Equuleus and Delphinus, sit just north of the celestial equator and are the starting point of this month's guide to the northern sky. They can be found between the brighter stars Altair in Aquila to the west and Enif in Pegasus to the east. Equuleus represents the little horse, distinct from the great winged horse Pegasus, further to the east. The faint 4th magnitude stars of the constellation form a slender triangle which points towards the south, almost directly away from the North Pole Star, Polaris.

To the north-west of Equuleus lies the other faint constellation, Delphinus, which from the northern hemisphere looks like a misshapen number 9. The two stars making up the western side of the '9' are Alpha Delphini, to the north, and Beta Delphini, to the south, and they have the rather unusual names of Sualocin and Rotanev respectively. We can thank the joking astronomer Nicolaus Venator for the strange alternative star names – he reversed his first name and surname and submitted them to the Palermo Star Catalogue. Gamma Delphini is the star to the east of Alpha and is a beautiful binary system composed of an orange giant star and a yellow-white companion. Thanks to our eyes' ability to play tricks on us, the contrast between the two makes the companion appear more blue-green to some people.

In stark comparison to the smaller and fainter Equuleus, to its east lies the much larger and brighter constellation of Pegasus, which represents a winged horse. Its brightest star, Enif, is visible as a moderately bright red star on the western end adjacent to Equuleus. The orange-red colour reveals that it is a cool star with a surface temperature of around 4,500 degrees and is fusing helium in its core into carbon and oxygen. It seems to be quite erratic in its behaviour, which is unusual for a star of this type: it has, on just a couple of occasions, brightened for a few minutes before fading again, perhaps as a result of giant flares.

Further east from Enif is the easily recognized square making up the greatest part of the constellation of Pegasus. The two stars marking the western side of the square are Scheat to the north, appearing red in colour, and Markab to the south. Just between the two and a little to the west is the star 51 Pegasi, which is home to the giant planet Bellerophon and was covered earlier in the chapter at page 214.

Hopping back to Enif again and moving a little to the north-west is a globular cluster called M15, the 15th object in Charles Messier's catalogue. It is one of the oldest clusters known and lies around 33,600 light years away. One of the most fascinating things about M15 is that half of its entire mass is squeezed inside a sphere with a diameter of 20 light years, which leads to the conclusion that the core is collapsing into a massive black hole. It has a magnitude of 6.1, so perfectly dark skies are essential to catch a glimpse of M15 with the naked eye, but binoculars will be needed to show it as a fuzzy star. To see individual stars a telescope aperture of at least 15cm is required.

On the other side of Equuleus and Delphinus is the much brighter star Altair, in Aquila, which forms the southern-most point of the asterism known as the Summer Triangle.

The other two points are made up of Vega in Lyra to the west and Deneb in Cygnus. Each star is prominent so the shape is very easy to spot. Deneb is the faintest of the three and marks the tail of Cygnus, and heading south back towards Delphinus is one of the swan's wings, marked by Epsilon Cygni, a yellow-orange star, and Zeta Cygni at the wingtip. Approximately halfway between the two stars and no more than a couple of degrees to the south lies one of the show-pieces of the northern summer skies, the Veil Nebula. The nebula is the remnant of a star which exploded as a super-nova about 7000 years ago and the shell of gas has been expanding ever since. It has two parts, an eastern and a western segment, and it is the former which is the easiest to observe, although it does require fantastically good atmos-pheric conditions to spot it. If a low-power, wide-field telescope is used to study the nebula, try adding a filter, which can greatly enhance what can be seen. The majority of light comes from oxygen molecules inside the nebula so, more specifically, an 'OIII' filter will make it pop into view. On the western side of Cygnus and moving out from the centre is the other wing, marked by Delta, Iota and Kappa Cygni. Kappa Cygni is famous for the meteor shower which is visible each August and that seems to emanate from a point not far from this star.

Back to Deneb and a little further to the east lies the North American Nebula, an emission nebula, so termed because it shines by emitting its own light, as opposed to a reflection nebula, which can be seen because it reflects light from nearby stars. The characteristic colour of an emission nebula is red but this is not seen when it is looked at with the naked eye because of the human eye's poor ability to perceive colour when light levels are low. Instead, it looks like a slightly richer part of the Milky Way, but with binoculars its whole outline can be seen, revealing the shape which gives it its name. A little further along the line of the Milky Way is M39, an open cluster of about thirty stars thought to be 800 light years away. It can just be seen with the naked eye but binoculars or a very low-power telescope are the best instruments to use because it covers a large area of sky slightly bigger than the apparent diameter of the Moon. The stars of the cluster are relatively bright in comparison to the stars in the Milky Way, which the cluster sits in front of, granting quite a beautiful view.

The stars of the main body of Cygnus point to the eastern side of the North Celestial Pole at a constellation called Cepheus, which looks somewhat like a wonky house with a pointed roof. Just to its south is a stunning star called Herschel's Garnet Star, with the most amazing deep red colour, and beyond it is another open cluster, designated IC1396, which covers an area of sky about four times the apparent diameter of the Moon. The cluster is visible to the naked eye but its surrounding nebulosity, referred to as the Elephant Trunk Nebula, requires the help of an OIII filter as for the Veil Nebula.

As August progresses, the stars set a little earlier every night, granting us a slightly different view of the night sky. This is because we mark a day as twenty-four hours but in reality it takes the Earth 23 hours 56 minutes and 4 seconds to spin once on its axis. It was way back in February/March that Leo was a prominent constellation but now it is lost below the western horizon even before midnight arrives. Interestingly, Regulus in Leo is one of the few bright stars in the northern hemisphere sky that can be occulted by the Moon, making for a really quite stunning spectacle, perhaps second only to a lunar occultation of the Pleiades star cluster in Taurus, and while these events are pretty rare, they are worth looking out for.

August: Southern Hemisphere Sky

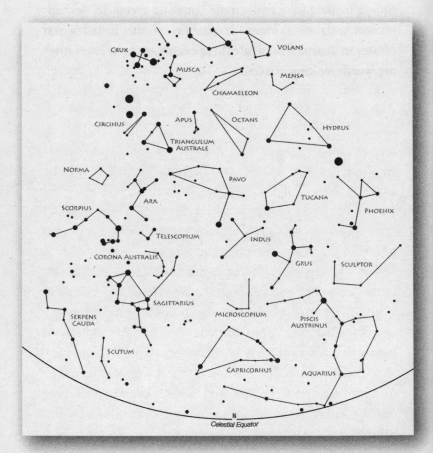

The plane of the solar system cuts across the Milky Way to the south of the celestial equator and it is along here that we can find the planets as they wander among the background stars. It is also along this line that we find the signs of the Zodiac, which are the constellations that the Sun passes through as the year progresses.

Capricorn is the easiest to find as it reaches its highest point for southern hemisphere observers around local midnight. It appears as a slightly deformed triangle with curved sides and its base runs almost parallel with the celestial equator. Its brightest star, Alpha Capricorni, sits at the eastern point of the triangle and to the naked eye it can be seen as a double star system, although in reality this is only an illusion as the two stars are unrelated and simply lie along the same line of sight. There is a meteor shower called the Alpha Capricornids whose radiant is close to Alpha Capricorni, although it is not an impressive shower, with just a few meteors visible per hour at peak.

The western point of Capricorn's triangle is marked by Delta Capricorni and between the two and about 5 degrees to the north is the location of a beautiful little planetary nebula called the Saturn Nebula. While it is visible with binoculars as a faint fuzzy star, a telescope with an aperture of 15cm or more is needed to see why it warrants its name. At magnifications of 200x and above, a couple of protrusions either side give the appearance of the rings around Saturn. Just off the south-western side of the triangle, a couple of degrees west of Zeta Capricorni, lies the globular cluster M30. Even small telescopes reveal a mottled fuzzy disc hinting at the individual stars that can start to be resolved through larger telescopes. William Herschel was the first to do this, in the early 1780s. In almost the same relative position on the eastern side of Capricorn lies another fainter cluster, M75, but at a distance of 67,500 light years it is one of the more remote globulars in Messier's list. Its diameter is 130 light years and with an estimated 100,000 stars it is probably one of the densest globular clusters known.

To the south of Capricorn is the rather faint and small constellation called Microscopium, which has no stars brighter than magnitude 4.9. As its name suggests, it represents a microscope and is one of a number of southern hemisphere constellations that depict scientific instruments. There are no great deep-sky objects in Microscopium; the most interesting is a spiral galaxy which is just visible through telescopes of 15cm and above but it is best seen with apertures of 25cm or more. Much brighter stars border Microscopium, with the stars of Sagittarius to the east and Fomalhaut in Piscis Austrinus to the west. The latter constellation looks like an upturned version of Capricorn but is a little flatter along its north–south axis.

A couple of faint constellations lie to the south of Microscopium, Indus directly to its south and Grus to the south-west. Indus looks like a flattened 'T' pointing towards the neighbouring constellation of Pavo to its east and is made up of stars no brighter than 3rd magnitude. The line marking the top of the 'T' is formed between the two stars Alpha Indi to the east and Delta Indi to the west. A line from Delta Indi extended to the south-west points to the constellation of Tucana's brightest star, Alpha Tucanae. The bright star between Delta Indi and Alpha Tucanae is Epsilon Indi, which shines at just under 5th magnitude. Given that it is the seventeenth-closest star to us, at a distance of 11.8 light years, it would be reasonable to think it to be among the brightest stars in the sky, but its low energy output of only 20 per cent that of the Sun means it is one of the intrinsically faintest stars visible to the naked eye.

Epsilon Indi is not only among the closest stars to us but is also the closest star to play host to a couple of orbiting brown dwarf stars. These are stars which are of such low mass that they have not managed to initiate full hydrogen fusion in the core. This leads to temperatures as low as 1000 degrees and there is some discussion about them forming the bridge in the gap between the evolution of stars and of planets.

To the south-east of Indus is Pavo, the peacock, and its nearest major star to Indus is its brightest. Beyond Alpha Pavonis along the same line and around 10 degrees away is a rather nice 6th magnitude globular cluster, NGC6752. Even small telescopes will start to reveal individual stars, which makes an attractive view with the 7th magnitude star HD177999 shining like a beacon in the foreground. Five degrees south of NGC6752 is the spiral galaxy NGC6744, which was host to a supermassive star that died and went supernova in 2005, showing that even deep space is subject to change. We looked at both NGC6744 and NGC6752 in the July sky guide, although they are a little better placed for observation this month.

The star visible to the west of NGC6744 is Delta Pavonis and it lies due south of Alpha Pavonis. Spectral studies of the star seem to show it has a higher abundance of heavier elements than is usual for similar stars, suggesting that it may well be orbited by a family of planets much like our Sun, although there is no other evidence for this yet. Continuing due west finds Beta Pavonis and Gamma Pavonis, a Sun-like star around 30 light years away.

To the south of Gamma Pavonis and roughly halfway between it and the South Celestial Pole is Nu Octantis, the brightest star in the constellation of Octans. It is classed as a giant star but is only three times the diameter of the Sun, and at a distance of 69 light years its low actual luminosity means it is not among the brightest in the sky. Nu has a companion star which orbits close at around 1 astronomical unit (the average Earth–Sun distance), completing one orbit in just under three years. The rest of Octans extends to the south-east and looks like a triangle, with Beta and Delta Octantis marking the other corners.

NINE

The Lives of Stars

LOOK UP AT the sky on any clear night and, if you are away from artificial lights, you can see up to 2000 stars at any one time, although that is a tiny fraction of the estimated 400 billion stars in our Milky Way. They seem to be there night after night, silently arching across the sky as the Earth turns under them, but the reality is, like most things, stars are born and will one day die. There will come a time many billions of years from now when every star we can see in the sky today, even the Sun, will have died and been replaced by a new generation.

Unlike many branches of science, we have a pretty good basic understanding of the life cycle of the stars, thanks to the vast number available for study. Learning about stellar evolution is not easy though, as most stars live for several billion years. It is similar to learning about the life cycle of great old trees which live for hundreds of years. The trick is that there are so many to study you do not have to watch one from start to end. Instead you can walk into a forest and see old trees, dead trees and young new ones, allowing you to draw a conclusion about the tree's life cycle. Astronomers use the same technique to learn how stars evolve by studying as many as possible, allowing them to infer the full cycle of a star's life.

When the Universe formed over 13 billion years ago, it was flooded with hydrogen atoms. Very slowly, over millions of years, gravity started to get a foothold, the atoms started to clump together and the more they clumped together, the more massive they were, and the stronger the gravitational pull became. This process continued as huge clouds of hydrogen gas slowly condensed into stars and, as they grew, the pressure in the core got higher and higher. When the pressure reached a certain point, hydrogen atoms started crashing together and, as they did, produced a tiny amount of heat and a tiny amount of light and it is this process that makes a star shine. The fusion process is actually part of a much bigger process known as nucleosynthesis, in which different elements are changed from one into another, driving the life of a star.

The first stars that formed were made almost entirely of hydrogen, but as they evolved, their core slowly changed from hydrogen to helium with a slowly shrinking shell of hydrogen on the outside. The process of nucleosynthesis persists throughout the life of the star as the pressure in the core increases and the helium atoms are fused into carbon. The same process continues over and over as other heavier elements are produced, such as oxygen and silicon – in fact, every heavy element in the Universe, including those that make up you and me, was originally produced inside the core of a star.

It takes billions of years for these stars to evolve and for the main portion of a star's life it sits quietly producing heat and light through fusion in its core. This leads to an interesting state where the pressure of energy production in the core, called thermonuclear pressure, pushing outwards tries to expand the star but the force of gravity produces a balancing force, halting the growth. The star is in adulthood and it remains in this state of equilibrium for billions of years.

The way in which stars evolve is pretty consistent and through understanding this we can look at any star in the sky and understand which stage it is at in its evolutionary process, and the way we do this is through a study of its spectrum. As we saw in Chapter 4, the only thing that determines the colour of light is its wavelength, which in the case of visible light is anything from 0.00039mm to 0.00075mm, and to put that in perspective, a human hair is about a hundred times thicker.

The relevance of the different wavelengths is that they determine how light interacts with materials, and in particular how much it is bent or refracted. A shorter wavelength is refracted more than a longer wavelength, which means that the shorter-wavelength violet light is bent more than the longer-wavelength red light. As a result of this, a beam of light will, when passed through a prism, split into its component colours because they are bent by different amounts. Nature demonstrates this beautifully in the rainbow, where water droplets act like prisms. Of course, astronomers do not rely on water droplets – instead we use instruments called spectroscopes, which attach to telescopes and act on the incoming light from distant stars in the same way as prisms.

In Chapter 5 I touched on the technique where, if a star's spectrum is looked at in fine detail, it is possible to see not only the individual colours but also a series of dark lines, called absorption lines, superimposed on the colourful spectrum. These lines are produced because a gas, present in the star, will absorb light passing through it and the type of gas present will determine the exact pattern of lines seen. Measuring the position and arrangement of lines, it is possible to work out what gases are present and therefore what a star is made of.

There is much more that can be understood from the spectra of stars and galaxies, but for now it is important to grasp how we can learn about distant stars just by looking at their light. You can do it for yourself, too, if you look at the colour of a star in the sky. A good example is Enif, at the south-western corner of Pegasus, or Altair, in the west of Aquila; by looking at their colour you can tell if they are a hot star or a relatively cool one. Imagine a workshop in which metal has to be heated to high temperatures. As the metal gets hot, its atoms start to give off light, initially shining with a red glow, then turning to yellow as the temperature increases, and on to white and ultimately blue. Simply by looking at the colour of the light given off from the metal, it is possible to tell roughly how hot it is, and in the same way it is possible to estimate the temperature of a star from its colour.

Stars differ in many ways other than colour; even their sizes span a huge range. There are some which are small and comparable in size to the Earth and others, like VY Canis Majoris, in the constellation Canis Major, which is thought to be around 2000 times the size of the Sun. If it were at the Sun's position it would be large enough to extend to a point just before the orbit of Saturn. Though VY Canis Majoris is probably one of the largest stars ever discovered, it is very inconspicuous in the sky and needs binoculars even to be glimpsed.

Whether a star is visible in the sky or not is determined primarily by two things: the amount of light it is actually producing and its distance from us. These two factors are combined to determine a star's absolute magnitude, which allows us to compare their real brightness rather than how they appear in the sky. This is done by calculating how bright the star would be at a distance of 10 parsecs (equal to 32.6 light years or 308.5 trillion kilometres, as described in Chapter 5), giving a basis for comparison. At a distance of 10 parsecs, in other words, the Sun's absolute magnitude is just +4.7, making it quite a faint star as we would see it in the sky, visible from dark sites only, whereas VY Canis Majoris is a mighty −9.4, which would make it brighter than Venus at its most luminous, even casting shadows here on Earth. As explained later in this chapter, a magnitude with a minus value is brighter than one with a positive value. The absolute magnitude scale is great for allowing us to compare the real brightness of objects, but to understand how bright they appear to us in the sky now, regardless of distance, we need another measure, called apparent magnitude.

The origins of the apparent magnitude scale come from Greek astronomers around 200 BC, when they divided the stars visible to the naked eye into six groups, with the brightest being assigned a value of one and the faintest a value of six. Each brightness value (or magnitude, to use the modern term) was estimated to be twice as bright as the next, so a 1st magnitude star was twice as bright as a 2nd magnitude star. The system was formalized around the middle of the nineteenth century by Norman Pogson, who defined a 1st magnitude star as 100 times brighter than a 6th magnitude star. This means the difference from one magnitude to the next is a multiplication by 2.51.

The system we use today is not limited to just the six magnitudes originally conceived by the Greeks, as the invention of the telescope revealed many more stars fainter than 6th magnitude. With the Hubble Space Telescope a distant galaxy was spotted at 30th magnitude, which places it at about one four-billionth the brightness of objects visible to the naked eye. We also have negative numbers in the modern magnitude scale to allow for objects brighter than 1st magnitude; for example, the Sun comes in at −26 and Venus at around −4.8 at its brightest. On maps of the sky, such as those in this book, the apparent magnitude scale for stars is represented by dots of different sizes, with fainter stars showing as smaller dots.

To complicate matters a little more, some stars vary in the amount of light we receive from them here on Earth, as a result of changeability either in the amount of light given off or in the amount that is blocked from reaching the Earth. These variable stars, as they are known, are common in the night sky but their presence is not particularly obvious; even the amount of light our Sun gives off varies, but only by about 0.1 per cent.

The first variable star was discovered in the seventeenth century and is known as Mira. It is a star about 300 light years away in the constellation of Cetus and is a binary star system made up of two components, a red giant star nearing the end of its life (Mira A) and a white dwarf star (Mira B). Mira A pulsates, leading to an increase and decrease in light output over a period of time, usually about eleven months. At its faintest it shines at around magnitude 10, making it undetectable by the naked eye, but at its maximum it can be seen shining at around 2nd magnitude just south of the celestial equator. The variability of Mira and the mechanism that causes this are not uncommon, and since its discovery several others have been identified and are now classified as Mira Variables. There are many other types of variable stars that actually change in their luminosity and it is a popular area of study for amateur astronomers. Any strange activity detected is then picked up by professional astronomers and studied in more detail.

Another reason for a change in the brightness of stars in the sky is their light being obscured by another orbiting companion star. The star Algol in Perseus is a good example of this; every three days its brightness dips from magnitude 2.1, when it is easily visible to the naked eye, down to 3.4. Its variable nature was first noticed in 1667 but it was almost a century later that the reason for this change was identified. It turned out that Algol was the first discovered example of a binary star system where the two component stars orbit in line of sight with the observer on Earth, so every few days one star eclipses the other, blocking a small amount of light from reaching the Earth and causing the star to dim a little. There are in fact two dips in brightness, a small drop only visible with instrumentation when the fainter of the two stars passes behind the brighter, and the main dip when the fainter star passes in front of the brighter one. There is actually a third star in the system but it is not responsible for any of the brightness changes we can see in the sky. The nature of this system was discovered by spectroscopic studies in which it was noticed that the star in the sky was wobbling back and forth as the smaller of the two tugged as it orbited, much like a hammer thrower spinning around. By studying this motion it is possible to calculate the mass of the two individual stars.

Aside from the spectroscopic binary stars like Algol there are many other stars in the sky that are part of binary or multiple star systems, many of which can be seen in amateur telescopes. Around 50 per cent of all stars in the sky have companion stars when studied but there are a few which are really stunning to look at.

Some of the stars in these binary systems orbit so close to each other that their proximity is quite destructive. If a red giant star orbits around a white dwarf, it is common for the red giant to spill material onto the white dwarf, whose mass increases slowly. Eventually the white dwarf reaches a critical mass and it explodes as a type 1a supernova, in the process ejecting its companion into space.

Violent explosions like this are not uncommon and are mainly associated with the death of a star. All stars die; even our Sun will one day, although this is not likely for another 5 billion years. Earlier in the chapter we looked at the nucleosynthesis process that transforms one element to another through fusion deep in the core of a star. Quite how a star ends its life is determined by its mass and the critical number is around nine times the mass of the Sun. When a star of less than nine solar masses has converted all the hydrogen in its core into helium it is left with a helium-rich core surrounded by a shell of hydrogen. The fusion process in the core now fades and, as a direct result of this, the force of gravity momentarily wins and the core is compressed. The compression of the core not only increases the temperature, from around 15 million degrees up to a staggering 100 million degrees, but the pressure too.

This is a key stage in the evolution of stars like the Sun as the increase in temperature and pressure starts to fuse helium in the core to carbon and the hydrogen shell into helium. As the fusion process picks up again, the outward thermonuclear pressure climbs, causing the star to swell up, increasing in size and energy output. In the case of the Sun, it is likely that it will increase in size so much that it will engulf the orbits of Mercury, Venus and perhaps even the Earth. As the star enters this new red giant phase of its life its increased energy production is spread over a much larger area, which results in a lower surface temperature and a shift in colour to the red part of the spectrum. Among the best-known examples of a red giant is Betelgeuse in the constellation of Orion, which shines with an unmistakable red light.

For most average-sized stars like the Sun this is as far as things go. Larger stars will continue to fuse carbon into heavier elements, but for the rest the repeated core contraction, temperature and pressure increase, followed by an increase in size and a subsequent reduction in temperature, will lead to the star slowly pulsing. The pulsations build up over time until they become so intense that the outer layers are eventually ejected into space and become a planetary nebula with just the core of the star remaining as a white dwarf.

A planetary nebula has nothing to do with planets although this term suggests a connection. They earned their name because their appearance through small telescopes is generally circular and planet-like. The reality of course is that they are the outer layers of a star expelled following the red giant phase. There are some beautiful examples of planetary nebulae within the grasp of modest amateur telescopes, such as the Ring Nebula in Lyra in the northern hemisphere and the Helix Nebula in Aquarius in the southern hemisphere. There are many more examples around the sky, some circular and others more dumbbell-shaped. It is thought the form is sculpted in some way by the magnetic field of the star, its rotation and also by the cloud's orientation in space.

Stars that are about nine times the mass of the Sun or

more suffer a rather more catastrophic end. Due to their larger mass the core compressions are more intense, with an increase in pressure and temperature that allows the fusion process to continue beyond carbon and through the heavier elements – neon, oxygen, silicon and iron – although each cycle lasts for a shorter period of time. Through these repeated cycles the outward pressure from fusion is helped in resisting the force of gravity trying to collapse the core by something called electron degeneracy pressure. Electrons are the tiny things that orbit around the nucleus of an atom and when they are packed closely together they try and move around more, generating pressure which, along with pressure from fusion, pushes against the force of gravity and resists further core compression.

Ultimately, though, this process is doomed to fail and when the core is finally fused into iron, further fusion will not produce any surplus energy so it is left to the pressure from the electrons to try and stop the collapse of the core. The iron core slowly gets bigger as the shell of silicon around it fuses into even more iron, until the mass of the core reaches about 1.4 times the mass of the Sun. At this mass, the pressure from electron degeneracy cannot overcome the crushing force of gravity and the core catastrophically collapses.

For a star that is less than twenty times the mass of the Sun another source of pressure stops a complete and total collapse and it is called neutron degeneracy pressure. A strange term but, effectively, during the collapse, electrons and protons (which along with neutrons make up atoms) crash together to form more neutrons and other particles. The neutrons are compressed so much that they exert an outward pressure and the core becomes one big neutron and further collapse is halted. Neutron stars are typically only a few tens of kilometres across but measure several times the mass of the Sun – in fact, a teaspoon of neutron star material would weigh the same as a few hundred big cathedrals. The collapsing outer layers rebound off the dense core and in an explosion which equates to letting off 100 million billion billion nuclear warheads they are ripped off into space in the blink of an eye. These are the type 2 supernovae, which are among the most violent events in the Universe.

The explosion of a supernova of this type gives off a phenomenal amount of energy – one star going supernova can outshine all the stars in a galaxy put together. Many examples of supernova remnants can be seen in the sky, such as the famous Veil Nebula in Cygnus, which is the result of a star that exploded about 7000 years ago.

If the original star is between twenty and forty times the mass of the Sun, it is set for an even more extreme fate since neutron degeneracy pressure is unable to oppose the collapse. The core of the star is compressed into an object even denser than a neutron star and known as a black hole. For the real giants among the stars, those over forty times the mass of the Sun, the star collapses completely into a black hole without ejecting any of the outer layers into space.

These strange and exotic objects rather defy common

sense as the entire core of the star is compressed down to an object, called a singularity, so small that it has no size, but is simply a point in space. Do not try and think about it too hard. The mass of a black hole is so high that even light, travelling at about 300,000 kilometres per second, is unable to escape so the black hole neither emits nor reflects any light, hence its 'blackness'. The boundary of the black hole is called the event horizon, which hints at the fact that any event beyond this point will never be seen by an observer in the outside Universe. Crazy stuff.

Black holes are not just the stuff of science fiction though, because we can detect their presence by measuring the movement of objects around them. By studying the spectrum of material being sucked into the black hole we can tell how fast it is moving and, from that, calculate the mass of the thing in the middle, the singularity. No black hole, by their very nature, has ever been directly observed, but the first strong candidate for one was found in the constellation of Cygnus, near the star Eta Cygni, and was named Cygnus X-1.

The planetary nebula and supernova stages mark the death of a star but also signal the start of another process lasting billions of years, in which the outer layers of the star will go on to form the next generation of stars. After the first generation of stars emerged following the creation of the Universe, the next generation formed out of material scattered throughout the galaxies from dying stars, including all the heavy elements created during the nuclear fusion process. By studying the chemical make-up of stars using spectroscopes it is possible to tell whether a star is one of the first, second or third generation of stars, which all have increasing amounts of heavy elements present.

In observing the stars we are not only learning about their lives, how they are formed and how they die, but gaining a real insight into how our Universe has evolved. Their study even leads us to understand where we have come from as every atom in our bodies has been synthesized inside the core of a star. So the next time you look up at a clear, dark sky, do not gloss over the stars to hunt down the planets and galaxies: take time to look at the stars – they are your heritage and they will not disappoint you.

September: Northern Hemisphere Sky

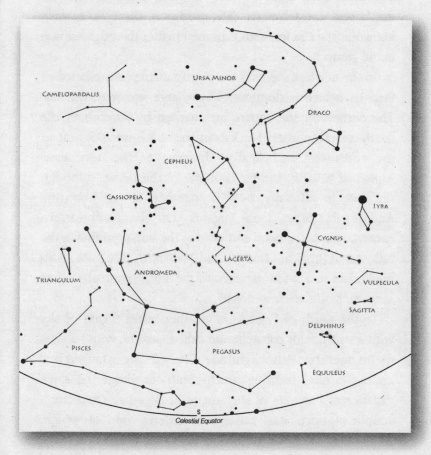

Sitting just to the north of the celestial equator in September are the stars that mark the extreme western end of Pisces. Forming what can only be described as an upturned classical pentagon when viewed from the northern hemisphere, the faint stars then extend out to the east, making it one of the largest constellations. The pentagon shape is actually known as the Circlet, with Gamma Piscium the brightest star in the group.

To the north of Pisces is the unmistakable constellation of Pegasus, which is dominated by a large square in the sky. The corners of the square are marked by Algenib in the south-eastern corner, Markab in the south-west, Scheat to the north-west and finally Alpheratz to the north-east. Alpheratz actually means 'shoulder of the horse', although the star is officially now a member of neighbouring Andromeda rather than Pegasus. Off the south-eastern corner extends the neck and head of the horse, ending at the red star Enif, the horse's muzzle. Scan the sky with binoculars just a few degrees to the north-west of Enif to spot the globular cluster M15.

To the north of Pegasus is a rather bland region of sky with a small, faint constellation called Lacerta, with its stars no brighter than 4th magnitude. The stars are arranged in a zig-zag fashion running north–south for about the same distance as one side of the square of Pegasus. There are a couple of open star clusters in Lacerta, one of which, NGC7243, is at the limit of visibility to the naked eye. It can be easily picked out with binoculars just to the west of Alpha Lacertae and 4 Lacertae at its northern edge.

Over to the east of Lacerta is the constellation of Andromeda curving up to the east from Alpheratz, the north-east corner star of Pegasus. At a point roughly halfway between Alpha Lacertae and Alpheratz is a small planetary nebula called the Blue Snowball Nebula. At magnitude 8.3 it will not be visible to the naked eye but telescopes of 15cm or more will reveal a slightly fuzzy disc with a hint of blue colour. Although there is some uncertainty over the distance, it is thought to be just under 6000 light years away and, if this is the case, that makes it just under a light year across.

Now getting higher in the sky is the well-known 'W' of Cassiopeia, which appears to the north-east of Lacerta. The stars marking the main shape of Cassiopeia are all more or less the same brightness and its most westerly star, Caph, is a useful pointer to an easy-to-find open-star cluster known as Caroline's Rose. Located just a few degrees to the south-west of the star, it was named after its discoverer, Caroline Herschel, in 1783. Binoculars or a low-power telescope will give the nicest view of this, while larger telescopes will reveal many more of its stars. Its age has been estimated at 1.6 billion years, which makes it one of the oldest clusters of its type.

Over to the west of Lacerta is the famous constellation of Cygnus, which looks like a great cross in the sky along the line of the Milky Way. Its brightest star, Deneb, is seen at the northernmost point and at the eastern wing tip is Zeta Cygni. Between Zeta Cygni and the slightly brighter star to its north-west is the Veil Nebula, one of the finest examples of a supernova remnant in the sky, although, unfortunately, one that is only visible to the naked eye under exceptionally dark skies.

Even further to the west is the distinctly bright star Vega in the constellation of Lyra. Off to its south-east is a faint parallelogram of stars and halfway between the two stars marking the southern side is the Ring Nebula, one of the best examples of a planetary nebula in the northern sky. The nebula is not visible to the naked eye; binoculars will reveal it as a fuzzy star but point a telescope at it to see it in its full glory, though the central star shown in photographs will only be seen through the largest of amateur telescopes.

Further north of Lacerta is the constellation of Cepheus, famous for being the home of Delta Cephei, the benchmark for Cepheid Variable stars. It looks like a slightly wonky house with its roof pointing to the north-east. The brightest star in Cepheus is Alderamin, found in its south-west corner. Just a few degrees to the south-east of Alderamin is Mu Cephei, otherwise known as Herschel's Garnet Star. It is one of the most deeply coloured stars in the sky, appearing as its name suggests a deep red. In reality it is a red super-giant star and is one of the largest stars in our galaxy. Just to its south is an open cluster of stars called IC1396, which is enshrouded in faint red nebulosity. It is difficult to see visually but looks incredible in photographs. Within the nebulosity is a dense and dark concentration of interstellar dust that gives the appearance of an elephant's head, hence its name, the Elephant Trunk Nebula.

Lying almost directly between the stars of Cepheus and

Caroline's Rose is another open cluster that was added to Charles Messier's catalogue as M52. It contains an estimated 200 stars, which through binoculars look like a faint misty patch of the Milky Way, but turn a small telescope on it and the stars of the cluster pop into view.

M52 lies almost an equal distance from two stars in Cepheus, Iota Cephei to the north and Delta Cephei to the south. Delta Cephei was the second variable star to be discovered that exhibited the period–luminosity relationship that is common among stars of this type. The variability is due to the star physically pulsating and at maximum brightness it will have swollen to forty times the diameter of the Sun before returning to its original size. The whole process lasts no more than 5.37 days and runs as regular as clockwork. Alderamin is the brightest star in Cepheus and is easily found to the north-west of Delta Cephei.

September: Southern Hemisphere Sky

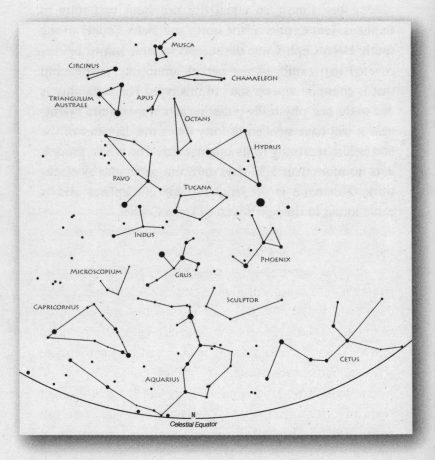

Aquarius, the water bearer, is a large but faint constellation well placed for observation in September and lies just south of the celestial equator. Alpha Aquarii is found almost on the celestial equator and is very slightly fainter than Beta Aquarii to its east. Due north of Beta Aquarii lies M2, a globular cluster that is great for smaller telescopes, although due to its huge distance of 37,500 light years larger apertures are needed to resolve the individual stars.

Over to the west of Alpha Aquarii is a large almost elliptical pattern of stars which represents the lower half of Aquarius' body and legs. The brightest star on the eastern side of the ellipse is called Skat and to its south lies the brightest star in this part of the sky, Fomalhaut in Piscis Austrinus. Imagine a triangle with these two stars marking the base, then the top of the triangle to the east is the location of the Helix Nebula, which at 650 light years away is the closest of all the known planetary nebulae. Its proximity means it appears large in our sky, just over half the size of the full moon, but regardless of that binoculars and small telescopes will show it only as a fuzzy blob. Larger telescopes will make it easier to see but the use of a technique called 'averted vision' will also help. This involves looking slightly to one side of the object and using the more sensitive parts of the retina.

The prominent 1st magnitude star Fomalhaut at 25 light years away is fairly close to us, which means its system can be studied in great detail. It has a surface temperature of around 8500 degrees and gives off fifteen times more light than the Sun. A red dwarf companion star orbits Fomalhaut at a distance of 1 light year, but in 2008 the discovery of planet Fomalhaut b was announced. It orbits at a distance of 115 astronomical units, taking at least 800 years to complete one orbit. This was also the first exoplanet to be directly imaged using the Hubble Space Telescope.

The area of sky to the west of Fomalhaut is devoid of bright stars and is where the constellation of Sculptor is found. It looks like a 'J' on its side marked out by mostly 4th magnitude stars. There are a handful of moderately bright galaxies in the western half of the constellation, notably the well-known Sculptor Galaxy in the north-western corner, although at magnitude 7 it needs a moderately sized telescope to appreciate it in its full glory. Phoenix lies to the south of Sculptor, with its brightest star Alpha Phoenicis, which looks distinctly yellow in colour and is the nearest bright star to Sculptor. Due east of Phoenix is Grus, which to the ancient Egyptians represented a flying crane, with Alnair as its brightest star to the south-east and Beta Gruis to its west, and together they represent the crane's feet. There is not much difference in brightness between the two stars but there is a stark contrast in the colours from the searing 13,500 degrees of the blue giant Alnair to the cooler red giant Beta Gruis at a modest 3400 degrees. Gamma Gruis at the northern end of Grus was once part of the neighbouring constellation, Piscis Austrinus, and marked the fish's tail, which is still reflected today in its alternative Arabic name, Al Dhanab, meaning 'the tail'.

Midway between Al Dhanab and Alpha Phoenicis is a faint cluster of four galaxies known as the Grus Quartet.

These four spiral galaxies, NGC7552, 7582, 7590 and 7599, lie at a distance of around 60 million light years and are all gravitationally interacting with each other. This can be seen from the bursts of star formation in two of the galaxies and tidal tails of stars and dust clouds being drawn out from the others. The galaxy cluster is quite faint though, and telescopes with apertures of at least 15cm will be needed to pick them up, with a little more detail being visible through larger instruments.

Moving further south from Grus leads to Tucana, a constellation that was introduced in the seventeenth century to celebrate the discovery of the Toucan in South America. It looks a bit like a diamond in the sky with a few extra stars thrown in to distort the shape a little. Alpha Tucanae is an orange giant star 200 light years away and is the brightest star in the constellation. It is found at the eastern point of the diamond, has a temperature of 4300 degrees, yet kicks out as much energy as over 400 of our Suns.

At the opposite end of the constellation is Beta Tucanae, which is a multiple star system made up from six individual stars. To the eye, two stars are visible, the brighter Beta-1 Tucanae and fainter Beta-3 Tucanae, which are apart by about 6 light years. Through a telescope, Beta-1 Tucanae separates out to become two stars, Beta-1 and Beta-2 Tucanae. Taking well over 150,000 years to complete an orbit of each other, the main pair are separated by over 1000 times the Earth–Sun distance. Each of the stars is then accompanied by a companion star, making this a sextuple system.

Over to the south-west of Tucana but still within its boundaries is NGC292, otherwise known as the Small Magellanic Cloud (SMC). It looks like a detached piece of the Milky Way to the naked eye yet at 200,000 light years is well beyond it. In the sky, it appears about six times the size of the full moon and by comparing this against its distance it is possible to calculate that it is around 7000 light years in diameter. As in many other galaxies, open clusters, globular clusters and nebulae can all be found in the SMC, many of which have entries in the New General Catalogue and can be picked out through large amateur telescopes with apertures of 25cm and above.

There are two globular clusters within the vicinity of the SMC: 47 Tucanae and NGC362. October's guide looks at 47 Tucanae in detail but NGC362 is sadly often overlooked because of the prominence of the other. It can be found slightly off the north-west edge of the SMC and at 6th magnitude is just detectable as a faint fuzzy star with the naked eye from a dark site. Binoculars will not show much more, a small telescope from 10cm upwards will start to reveal more detail, but a 15cm telescope or larger is needed to be able to start resolving some of the individual stars. It lies at a distance of 27,700 light years and is 130 light years in diameter.

TEN

The Realm of the Galaxies

W E LIVE ON A planet orbiting around a fairly average
star which itself orbits around the centre of something we call the Milky Way. Nearly every single object that
can be seen in the night sky is also a member of this vast
galactic family, which we have named after its ghostly, milky
appearance. It is like a giant island in the never-ending
emptiness of space and is home to most objects that light up
the night-time sky. From the 4000 or so stars that can be
seen by the naked eye, to the planets, star clusters and gas
clouds, they are the objects that, bound together by the
force of gravity, make up our galaxy. Yet there are a couple
of objects in the sky, mere smudges to look at, which hint
that the Milky Way is not alone.

The reality is that the Universe is peppered with billions
of galaxies of different shapes and sizes and with varying
numbers of stars in each; indeed, it is often said that there
are more stars in the Universe than there are grains of sand

on the Earth. Perhaps one of the most exciting aspects of astronomy is that not only are the furthest corners of the cosmos being probed by studying the galaxies, but also the depths of time. The nearest major galaxy to our own, the Andromeda Galaxy, is visible to the naked eye in the constellation of Andromeda and lies a whopping 2.3 million light years away. That is nothing, though, compared to the most distant object so far discovered, a galaxy with the catchy name UDFj-39546284 a mind-boggling 13.2 billion light years away – the light forming the image we see today left that galaxy some 8 billion years before our Solar System began to form.

Before looking at some of the other galaxies scattered throughout the Universe, it is interesting to examine first how our view of the Milky Way has changed over the years. When humankind first looked at the sky the view was unimpeded by artificial lighting and the spectacle would have been truly stunning, achievable today only from some of the most remote places on the planet. Our ancestors would have seen the band of stars defining the Milky Way with amazing clarity, yet it was not until the invention of the telescope and the curiosity of Galileo in probing the band of light that things started to change. He found that, under magnification, the light separated out into thousands of individual stars.

Not much changed until another astronomer, William Herschel, who was working at his own observatory in England, turned one of his large telescopes on the Milky Way to try and measure the distance to as many stars as possible. Making the rather rash, and incorrect, assumption that all stars give off the same amount of light, he estimated their distance based on their apparent brightness in the sky, fainter ones therefore being further away than brighter ones. We now know that stars vary considerably in the amount of light they give off so his estimates would have been quite wrong even though he was simply working on a comparison of distances with each other rather than actual distances from us. He drew the conclusion that we were located inside a giant disc of stars, with the Milky Way representing the plane of the disc, a pretty accurate view.

Other than Herschel's disc-shaped view of our galaxy, very little was known about its actual size and shape until 1914, when another astronomer, Harlow Shapley, started to study clusters of stars with the 1.5m reflecting telescope at Mount Wilson Observatory in California. He found that some of those under observation seemed to contain a type of variable star whose actual light output was directly linked to how long it took the star to change from minimum to maximum brightness. By observing these very special Cepheid Variable stars in the distant clusters, he could time how long it took for them to change in brightness and therefore deduce how much light they really gave off. Comparing this to how bright they seemed in the sky would allow him to calculate their distance and, hence, the distance to the cluster.

When Shapley plotted the positions of some of the clusters, a remarkable picture emerged. Their distribution seemed to be centred on a point a staggering 60,000 light years away, and the galaxy itself appeared to be about 300,000 light years in diameter. We now know that the diameter is in fact about a third of Shapley's figure, at around 100,000 light years, and the galactic centre is around 30,000 light years away in the direction of Sagittarius.

Since Shapley's studies, a lot more has been learnt about the shape of the Milky Way, primarily by using radio telescopes to study the locations of clouds of hydrogen gas out of which stars form. By mapping their distance and position a picture emerges of a galaxy shaped like a flattened disc with spiral arms emanating from a bar which runs across a central bulging nucleus. The name for this type of galaxy is not surprisingly a barred spiral galaxy. A good analogy for the appearance of the Milky Way from the side is two fried eggs stuck back to back with the white representing the plane along which the spiral arms exist and the yolk representing the central bulge. More detailed studies of the motion of gas clouds at the centre of the galaxy show that there is a central object of very large mass. By measuring the speed of the orbiting gas clouds it is possible to determine the mass of the object and it is now believed that a large supermassive black hole lurks there.

The Milky Way is not alone though, as there are an estimated 170 billion galaxies in the observable Universe. The first recorded observation of another galaxy was as far back as the tenth century, by a Persian astronomer, Abd al-Rahman al-Sufi, who had noted a 'small cloud', which we now call the Andromeda Galaxy. Without the magnifying power of a telescope its true nature was not known – in fact, for many years it was referred to as a 'nebula' (Latin for 'cloud') due to its fuzzy appearance. In the centuries that followed there were many theories as to the origin of these fuzzy blobs, some of which were not all that wrong, including the English astronomer and mathematician Thomas Wright's, who in 1750 suggested the tiny faint smudges were separate 'Milky Ways'!

The real breakthrough came with the invention of the spectroscope, used in 1912 by the American astronomer Vesto Slipher to study the nebulae, which appeared to be spiral in shape. He was trying to learn what they were made of but instead discovered that they seemed to have a high red shift, which is seen as an apparent shifting of certain features in the spectrum towards the red end. This showed that the objects were heading away from us at a great speed, faster than the speed needed to escape the gravitational pull of the Milky Way, so they cannot have been gravitationally bound together. The conclusion was simple: they were a long way outside our galaxy. A figure was finally put on the distance to the brightest of the spiral nebulae, the Andromeda Galaxy, by the American Astronomer Edwin Hubble, who was using a giant 2.5m telescope. By trying to resolve individual stars in our galactic neighbourhood, he found Cepheid Variable stars, which were used to help determine distance. This led to the currently accepted figure of 2.5 million light years.

When Edwin Hubble started to study galaxies he realized that they fell into three broad categories: spirals, ellipticals and irregulars. Each category is then subdivided further, with spiral galaxies defined by the tightness of the spiral arms and whether they have a central bar, and elliptical galaxies by their shape from spherical to ellipsoid (shaped like an egg). The irregular galaxy category is a curious one and is more of a catch-all group for galaxies that do not fall into the other categories. Which one a galaxy belongs to is determined purely by its physical appearance and not by any other features, such as star formation rate, for example.

The shape of a galaxy seems in no way related to a stage in its evolution; generally, once a galaxy forms, it broadly retains its shape unless it happens to collide with another galaxy. The formation process that determines the shape of a galaxy is still very much under debate, but we do know

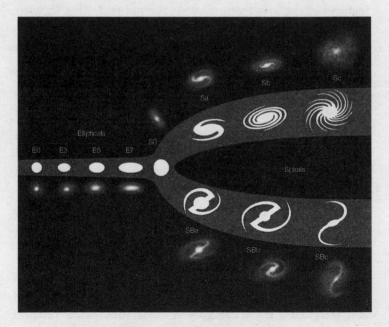

Hubble Galaxy Classification

that all galaxies, whether elliptical or spiral, emerged over millions of years from tiny fluctuations in the distribution of matter after the Big Bang. It is thought that the tiny fluctuations grew slowly and eventually small proto-galaxies started to appear. Over time they clumped together, creating the galaxies we see today, and the gas slowly formed into the first generation of stars.

The theory that galaxies formed out of the variation in matter distribution after the Big Bang works well, but it does not describe why we see different types of galaxies. The spiral galaxies, for example, differ quite markedly from the large elliptical galaxies, in that they are really quite thin relative to their diameter and rotate quickly, and generally we see a significant amount of star formation going on in them. The elliptical galaxies, on the other hand, are much larger and do not rotate, and there is barely any star formation going on. To understand why we see different galaxies we need to look at their formation and interaction in a little more detail.

Shortly after the Big Bang, the tiny fluctuations grew, leading to massive haloes of so-called dark matter, which is material that neither emits nor absorbs electromagnetic radiation. The only way it can be detected is through its gravitational interaction with other matter and light. The proto-galaxies formed out of the dark matter haloes and, over time, the smaller galaxies merged to form larger galaxies with the dark matter haloes remaining around the outside. We can see this distribution of dark matter haloes even around today's galaxies. The gas content of the young massive galaxies quickly contracts under the force of gravity and it is this initial motion which starts the galaxy spinning. This rotating motion tends to force the material outwards, producing a thin, disc-shaped galaxy. For some unknown reason, the contraction seems to cease, perhaps because of the rotational movement of the galaxy or maybe due to the gravitational pull from the dark matter, but for now the cause remains a mystery.

The classic spiral galaxy seems to form the greater proportion of galaxies and in appearance these resemble a flat rotating disc that is home to the spiral arms which emanate from a central nucleus. Surrounding the galaxy, along with the invisible dark matter halo, is a nearly spherical halo of stars which is made up almost entirely of dense globular star clusters. One of the long-standing mysteries of spiral galaxies is just how they retain the spiral structure. It is not possible for the stars to be arranged in a spiral structure because over time the arms would wind up tighter and tighter until they eventually dissipated. Studies of the motions of stars in the spiral arms actually show the more distant stars rotating faster than expected, which leads to two possible explanations.

The older of the two theories suggests a phenomenon known as density waves, which rotate around the centre of spiral galaxies. They can be thought of like a traffic jam on a motorway: individual cars move through it but the traffic jam stays where it is and does not move. In the same way, stars move through the wave and so observationally you would see a different set of stars in the arms at different times as the wave continues on its orbit. When the wave

meets gas, it compresses it, leading to a burst of star formation and the appearance of young, hot and luminous stars
that make the arms stand out from the surrounding galaxy.
The other, rather less popular, idea suggests that the
explosion of stars and solar wind being emitted from all
stars generates shockwaves which continue on around the
galaxy. As the shockwaves interact with gas, they compress
it, leading to a burst of star formation in the same way as in the
density wave model. M33 is a great example of a spiral galaxy
in the northern skies during October and is just visible to the
naked eye under exceptionally good conditions between the
Andromeda Galaxy and the stars of Aries.

A slight variation of the spiral galaxy is the barred spiral,
which retains the spiral arms feature but, instead of protruding
from the nucleus, they sweep out from a bar which itself
extends through the nucleus. Interestingly, surveys have shown
that there are more barred spiral galaxies than pure spirals. The
widespread presence of the barred spirals suggests that
the existence of the bar may well be an evolutionary stage in
the life of a spiral galaxy, and it is thought that these may be the
younger relatives of the fully grown spirals. Orbital resonance
is the mechanism that may be the driving force behind the
strange feature where stars and gas in orbit around the nucleus
exert a gravitational force on each other. Eventually their orbits
become gravitationally synchronized, leading to the formation
of the bar. This process may also encourage the birth of new
stars in and around the core of the galaxy until the bar reaches
such a mass that it becomes unstable and the galaxy evolves
into a spiral. There are a number of examples of barred spiral

galaxies around the sky but NGC55 is one of the brighter ones, although appearing edge-on to us. Like most galaxies it is not visible to the naked eye, but a beginner's telescope will hint at some detail. It is possible that the Small Magellanic Cloud in Tucana was once a barred spiral galaxy that has been distorted by the immense gravity of the Milky Way.

The other main class of galaxy is the elliptical, whose appearance varies considerably from almost spherical galaxies at one end through to nearly cigar-shaped at the other. The stars in the elliptical galaxies travel in random motion, unlike those in spiral galaxies, which orbit a central nucleus. This is seen clearly to be the case in M32, one of the companion galaxies of the Andromeda Galaxy.

Typical elliptical galaxies are made up of old, low-mass stars and have almost negligible quantities of gas, which means star formation is rare. The absence of star formation suggests that elliptical galaxies are old, but surprisingly they seem to be few and far between in the early Universe. We can tell this because it takes time for light to travel the great distances between the galaxies, so looking at distant ones means we are looking back in time. This leads us to the conclusion that the giant ellipticals are actually the result of collisions between, or mergers of, two or more galaxies of equal mass, rather than having originally been formed in that shape.

Given the incredible distance between the galaxies it is hard to believe that they can ever get close enough to merge. Our own galaxy, the Milky Way, is over 2 million light years away from its nearest major galactic neighbour, the Andromeda Galaxy, yet even they are heading towards each other. There is still some uncertainty as to whether they will actually collide, but if they do, it could happen in as little as 5 billion years from now. This all sounds pretty scary stuff but the reality of galactic collisions and mergers is not quite as terrible as it seems. Without you even realizing it, the Milky Way is in the process of devouring a smaller galaxy called the Sagittarius Dwarf Elliptical Galaxy. Typically, when galaxies collide they have a relative impact speed of around 500km per second, yet very little real damage is done. Because of the distance between the stars, it is probably more accurate to call it a merger rather than a collision; indeed there are very few if any impacts. To put that in context, if the average star were scaled down to the size of a golf ball and placed in the centre of London, then the next nearest star would be at Zurich in Switzerland and the chances of the two of them colliding would be negligible.

The series of events during a merger is very much dependent on how the process unfolds, but we do know collisions occur as there are many fine examples of them in the sky such as the Large and Small Magellanic Clouds in the southern hemisphere sky, which are now thought to have collided with the Milky Way billions of years ago. An example in the northern sky is the Whirlpool Galaxy, found in Canes Venatici. The system is stunning through a telescope, which reveals two galaxies in the process of colliding. It can just be seen through binoculars, although detail in the spiral arms is only revealed in telescopes.

If collisions like this take place between two equal-mass spiral galaxies such as the Milky Way and Andromeda, then before getting too close the stars will orbit around the respective nuclei in a fairly orderly fashion. As the galaxies get closer and start to merge, the gravitational effects throw this all into disarray and the stars are disturbed in their orbits, sending them off on different paths. Interestingly, this is what we see in elliptical galaxies, which supports the theory that ellipticals are the result of galaxy mergers. The two galaxies will for a short time maintain their original motion so they pass through each other, sending the stars into random orbits and disturbing the gas, stirring up a burst of star formation. The shape of the galaxies will change dramatically, with some stars and gas being ejected into space. If the galaxies have insufficient momentum then the force of gravity will pull them back together again and the process continues until the galaxies eventually merge. On rare occasions the galaxies will be massive enough and travelling fast enough for them to continue on their way and not merge into one. Instead, the transient gravitational interaction will simply change their appearance for ever.

At a local level it seems galaxies are sparsely distributed – after all, the 2.3 million light years or 21 trillion kilometres to Andromeda does seem quite a distance. Looking further afield at the wider cosmos, galaxies are clustered together into gravitationally bound groups. There is actually a difference between a galaxy group and a cluster but it is chiefly one of size, with galaxy groups having up to fifty members and measuring around 6 million light years across. Clusters are larger than this but there is no other real differentiation between them and groups.

Our own galaxy is part of a group of galaxies called the Local Group, of which there are around forty known members. The two largest members are the Milky Way and the Andromeda Galaxy, and they are joined by a host of smaller galaxies, including the Small and Large Magellanic Clouds, which lie at a distance of 200,000 light years and 160,000 light years respectively. They can only be seen in the southern hemisphere sky and appear like disconnected patches of the Milky Way; and are small in comparison to many other galaxies, including the Milky Way, measuring just 7000 and 14,000 light years across.

Beyond the Local Group are more clusters of galaxies, such as the Virgo Cluster – which contains up to 2000 galaxies and includes the giant elliptical galaxy M87.

The Virgo Cluster, the Local Group and almost a hundred other galaxies and clusters are collectively known as the Virgo Supercluster. A galaxy supercluster is the largest structure in the Universe, and in the case of our own Virgo Supercluster measures 100 million light years from side to side. Interestingly, superclusters, of which there are about a million in the observable universe, are not gravitationally bound. Their existence is believed to represent the large-scale fluctuations in the Universe following the Big Bang and between them are vast expanses of empty space where few if any galaxies exist.

There is one final type of galaxy to be considered and it has been left to the end of the chapter due to its strange and exotic nature. These are the galaxies with something quite unusual at their core called the active galactic nuclei. These 'active' galaxies have a higher than usual output of energy at their core, be it in visible light, X-rays, radio waves or any other portion of the electromagnetic spectrum. It is thought the source of the energy is the accretion of matter around a supermassive black hole at the centre of the galaxy. The gravitational pull of the black holes is so high that they drag matter towards them at high speeds. As the matter gets nearer it forms an accretion disc which spins faster and faster the closer it gets to the black hole's event horizon (the point where even light does not travel fast enough to escape). As matter is accelerated to higher speeds it heats up and emits the intense quantities of energy observed. Another feature of some active galaxies is jets of radiation (relativistic jets) that extend out over millions of kilometres and, while the exact cause of these is unknown, they do seem to contribute significantly to the overall brightness.

There are a few different types of active galaxy which are divided into two sub-categories: those that emit highly in the radio wavelengths, called 'radio loud', and those that do not, called, unsurprisingly, 'radio quiet'. The key difference between the two is the emission from the jet that is dominant in radio wavelengths in the radio loud group. Within the radio quiet group is the best-known type of active galaxy, the quasar, which is short for 'quasi-stellar object', and Seyfert galaxies, which were the first active galaxies to be identified. Within the radio loud group are the blazars, which are the most luminous of all. All of them have supermassive black holes in the centre, surrounded by an accretion disc, and the so-called relativistic jets, but they differ because of their orientation in space relative to our vantage point here on Earth.

The Seyfert galaxies, like M77 in Cetus, which is 47 million light years away, are the closest, and in the case of M77 can be seen visually as a barred spiral galaxy in amateur telescopes. While they are the nearest of all active galaxies, it seems that the relativistic jet is not pointing towards the Earth, so while they do have bright cores they do not appear to be among the most energetic. The quasars are generally seen to be more energetic than the Seyfert galaxies but are much more distant. When they were discovered, they appeared like stars visually but seemed to be in exactly the same part of the sky as a strong radio source. The beam of a quasar is thought to be pointing more towards the Earth than is the case with Seyferts, but still not directly at us. The closest quasar is an object called 3C273 at a distance of around 2.5 billion light years in Virgo, but a good-sized amateur telescope is needed to see it. The blazar is the most energetic of all with significant observed emissions of radiation, particularly in the radio spectrum. They seem to be hosted inside giant elliptical galaxies and it is thought the beams are pointing directly at us so we see the characteristic intense amounts of energy coming from a compact source.

The family of galaxies around the Universe is a diverse one and, while space is to all intents empty, these giant islands light up the darkness. It is fascinating that they are all generally home to the same objects yet their appearance in the night sky varies wildly, making them a great target for amateur observations.

October: Northern Hemisphere Sky

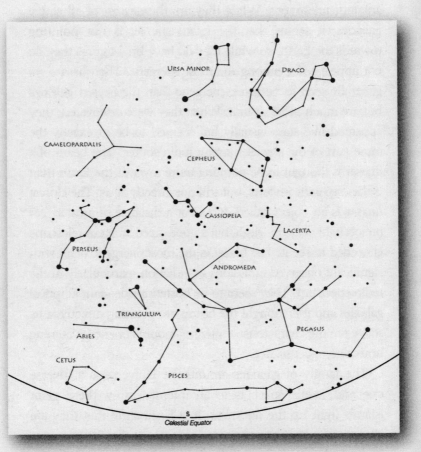

Pisces is very well placed in the northern sky during October and sits just north of the celestial equator. To the west can be seen the Circlet pattern of stars, which actually looks more like a pentagon than a circle. The first bright star to its east is Alpha Piscium, which is the third-brightest star in the whole constellation. The rest of Pisces forms a large V-shape, with Alpha Piscium at its tip, that points to the south-east.

To the west of Pisces is the unmistakable Square of Pegasus. The square itself is just a portion of the constellation, which represents the winged horse, but the four stars marking the corners of the square are a great signpost to the October sky. The south-east corner star, known as Algenib, is a striking blue star with a surface temperature of 21,500 degrees, but of real interest is NGC7814, just a few degrees off to its north-west. NGC7814 is a beautiful edge-on spiral galaxy but a little faint at 10th magnitude, although even a small telescope will reveal its long, thin shape. Telescopes with an aperture of 15cm or more will start to show the dark dust lanes against the faint light of the galaxy.

The north-east corner star of the Square of Pegasus, Alpheratz, actually belongs to the neighbouring constellation of Andromeda, found just to its east. Continue along the line of stars from Alpheratz to the east, past Delta Andromedae, then turn slightly north to find Mirach, a cool red giant star. Now turn further north and locate Mu Andromedae, then a little further on the slightly fainter Nu Andromedae, before shifting your gaze fractionally to the west to find a faint fuzzy blob. This is the famous Andromeda Galaxy and from dark locations it is quite easy to spot with the naked eye. It is the nearest major galaxy to our own and lies around 2.3 million light years away, although, in contrast to the motion of NGC7814, which is heading away, it is actually heading towards us. The Andromeda Galaxy is an easy target for the naked eye under dark skies and with binoculars, but telescopes will reveal dark dust lanes. Small telescopes will even reveal its two companion galaxies, M32 and M110, much like the companions of the Milky Way, the Large and Small Magellanic Clouds that can be seen in the southern hemisphere sky.

Starting back at Mirach again (the bright star to the north-east of Alpheratz), find the orange star to its south-east, which is Hamal in Aries. This one is easy to spot as it is prominent in an otherwise bland area of sky. Between the two is a faint constellation called Triangulum, in the shape of a thin triangle and pointing to the west. A little further off to the west of the triangle itself and almost halfway between Mirach and Hamal is M33. At a distance of almost 3 million light years it is the most distant object that can be seen with the naked eye under exceptionally dark conditions. Binoculars will reveal it as a large fuzzy blob but it takes a telescope with an aperture of at least 15cm to start to reveal a hint of the spiral structure.

The prominent star to the north of Triangulum is the most easterly star of Andromeda, known as Almach. Further east of Almach is another bright star, by the name of Algol, in Perseus, and about a quarter of the way towards it is another fine example of an edge-on spiral galaxy, NGC891. At 9th magnitude it requires a telescope to be seen in any detail, but even a small one will show a thin needle-like smudge of light. Larger telescopes will reveal a dark dust lane along its equator. Studies in infra-red light suggest the galaxy has a bar similar to the Milky Way but with its edge-on orientation it is not directly visible to us.

To the north-west of Almach are a couple of fainter stars separated by no more than a couple of degrees: Upsilon Persei (orange) and Phi Persei (white-blue). To the north of Phi Persei by a fraction of a degree is one of the faintest planetary nebula, M76, otherwise known as the Little Dumbbell Nebula because it resembles a smaller version of the famous Dumbbell Nebula found in Vulpecula.

Further north from Andromeda lies an easy-to-recognize group of stars called Cassiopeia, which looks like a giant celestial 'W'. Caph, the star at the western end of the 'W', lies 54 light years away and is thirty times more luminous than the Sun. The next star to the east in the shape that makes up the 'W' is a yellow star by the name of Shedar. It looks to be of a similar brightness to Caph and therefore it would seem safe to assume it is at roughly the same distance but it is actually over four times further away. This illusion is a result of Shedar's greater size and luminosity – it is 855 times brighter than our Sun. Just to the east of Shedar is the emission nebula known as the Pacman Nebula, which, at magnitude 7.4, is only visible with optical aids. A telescope aperture of at least 10cm should reveal it nicely, but more detail, such as the dark dust lane cutting into the nebula, will only be seen in large instruments. To the eye, the nebula appears grey-green, but in images of the area a beautiful red light is seen, caused by energy from nearby stars exciting the hydrogen-gas atoms in the cloud and making them glow. Those same hydrogen atoms were created at the birth of the Universe.

Moving further along Cassiopeia to the east finds the stars Navi and Ruchbah, and scanning the sky with binoculars to the east of Ruchbah by 1.5 degrees reveals M103, a 7th magnitude open cluster. It is not so easy to spot with a telescope as the larger aperture, and therefore greater light-gathering power, also reveals more background Milky Way stars, making the cluster much less obvious. It lies about 7000 light years away and, like many open clusters, has a small number of stars – forty in this case.

October: Southern Hemisphere Sky

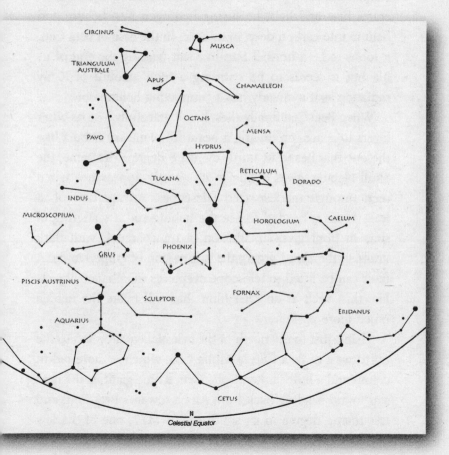

The constellation of Cetus straddles the celestial equator and is the starting point for October's southern sky guide. Its brightest star, Deneb Kaitos, also known as Beta Ceti, is prominent just 18 degrees south of the equator and is easy to spot due to its yellow-orange colour. Like all stars it generates heat and light by fusing hydrogen into helium and helium into carbon deep in its core. In the case of Beta Ceti, it looks to be a normal Sun-like star nearing the end of its life but it seems to be emitting a high amount of X-ray radiation as if it already has a contracting helium core.

When Beta Ceti finally dies, it will gracefully lose its outer layers to space, producing a beautiful planetary nebula like the one that lies to its north by just 8 degrees. Its name, the Skull Nebula, gives a clue to its ghostly appearance when seen through modest-sized telescopes. An aperture of at least 15cm is needed to see the nebula and a scattering of stars in front gives an illusion of transparency, with dark knots in the cloud giving the impression of empty voids. A filter can be fitted to telescope eyepieces to enhance objects like this, such as an OIII filter that will make the nebula much more prominent.

Visible just to the north of the celestial equator and to the north-west of Beta Ceti is Alpha Ceti, which is more red in colour and a little fainter. Delta Ceti, a blue giant, is the next star found heading back from Alpha towards Beta Ceti, and less than a degree to its south-east is M77, one of the few Seyfert galaxies visible to amateur astronomers. It appears as a 9th magnitude barred spiral galaxy face-on to us. Heading from Alpha Ceti through Delta Ceti and on the way to Beta

Ceti is a famous star called Mira. It is a red giant whose light output varies as the star pulsates over a period of 332 days. In that time, it brightens to a maximum magnitude of 3.5 before slowly fading from view to magnitude 9.5. It has a temperature of around 2000 degrees, making it one of the coolest stars in the sky. The name Mira, which means wonderful, is now used to describe variable stars of this type and there are about 6000 of them across the whole sky.

Starting back at Beta Ceti again, the next bright star to the south and roughly as bright is Alpha Phoenicis, or Ankaa. It is a yellow-orange star about 77 light years away and is joined to its south-west by Beta Phoenicis, which is a little fainter. Between these two stars in Phoenix and Beta Ceti is Alpha Sculptoris, the most westerly of the bright stars in Sculptor and a blue giant star around four times the diameter of the Sun. Sitting almost at the centre of a triangle made up from Alpha Sculptoris and Alpha and Beta Phoenicis to the south is a beautiful example of a spiral galaxy, NGC300.

NGC300 covers about the same area of the sky as the full moon and at a distance of just over 6 million light years is one of the closest galaxies to our Local Group of galaxies. It was once thought to be a member of the Sculptor group of galaxies but the cluster lies some 4 million light years further on. At magnitude 8, it is not visible to the naked eye but telescopes with an aperture of around 10cm will show it as a fuzzy disc, with larger instruments starting to reveal detail in the disc, but the spiral arms only become prominent in telescopes of 25cm or more. To the west of NGC300 by around 10 degrees is NGC55, an example of an edge-on barred spiral galaxy, lying 7 million light years away.

South of Phoenix and slightly to the west is the bright star Achernar, the ninth-brightest star in the sky and the brightest star in the constellation of Eridanus, which depicts a river. It marks the end of the river and is a giant blue star 144 light years away with a temperature in excess of 15,000 degrees. It spins on its axis at a speed of 225km per second, which causes the star to bulge out around its equator so its diameter is less from pole to pole. The remainder of the stars in Eridanus extend to the north past the constellations of Horologium, Fornax and Caelum before almost reaching the celestial equator near Rigel, the brightest star in the constellation of Orion.

To the south of Achernar is a large celestial triangle with its northernmost tip pointing just to the west of it. The triangle is the constellation of Hydrus and is marked by Alpha Hydri at its northernmost tip, Beta Hydri at the eastern tip and Gamma Hydri to the west. The brightest of the stars is Beta Hydri, which lies 24 light years away and is one of the most Sun-like stars in the night sky, though it is probably a little older and a little further down its evolutionary path than our star at an estimated 6 billion years old.

Along the line between Beta and Alpha Hydri is what seems to be a slightly detached portion of the Milky Way. It is actually one of our four closest satellite galaxies, the Small Magellanic Cloud (or SMC), the other three being the Sagittarius Dwarf Elliptical Galaxy, the Canis Major Dwarf Galaxy and the Large Magellanic Cloud. Just off the eastern edge of the SMC is a 4th magnitude star that looks a little hazy to the naked eye. Binoculars simply make the fuzzy star look bigger and there is an increase in brightness towards the core. Using a telescope with at least a 10cm aperture will start to turn the haze into individual stars, revealing the globular cluster 47 Tucanae in all its glory. The cluster extends over an area of sky equivalent to the full moon which in reality is 120 light years across. In that area of space there are around a million stars, in comparison to an equivalent region of space around our own neighbourhood where there are an estimated 15,000 stars.

ELEVEN

The Evolution of the Universe

W E LIVE IN a weird yet wonderful Universe; take black holes as an example. Essentially a black hole is the thing that gets left behind after a huge star explodes at the end of its life. As we've seen, the forces can be so extreme during the final part of the process that the core of the star gets compressed to a tiny, tiny thing called a singularity. Nothing strange about that you might think, but understand that the singularity is so small it has no size, yet is so heavy no scale has large enough numbers to describe it – its density is infinite. Perhaps it is stranger than you first thought.

As we continue to study the Universe at small and large scale with ever-more sensitive equipment we find that things just do not make sense any more to our 'everyday logic'. Black holes are a fine example of this, as is the atom. Atoms are made up of tiny balls of energy with some at their centre (or nucleus) called protons and neutrons and others whizzing around the nucleus called electrons. These orbit the nucleus just like the planets orbiting the Sun, but that is where the similarities end, as planets usually stay in their orbit but the atom's electrons move from one orbit to another as energy is absorbed or emitted. Here is the really strange bit: when electrons move between orbits, they do not move like you or I do from one place to another – they actually disappear from one orbit and suddenly reappear in another without travelling through the space between. It is not just the small scale of singularities or atomic structure which defies common sense: the Universe itself, its beginning, its ultimate fate and the very structure of space have challenged scientists for hundreds of years. One of the biggest unanswered questions in science today is: how did it all begin?

As we saw in Chapter 1, our understanding of the creation of the Universe dates back to the 1920s, when Edwin Hubble spotted something peculiar while he was looking at the spectra of distant galaxies. Among the first things he noticed were the tell-tale signs of the existence of several types of gas, through the presence of absorption lines against the spectrum. But there was something different about them: instead of being in their usual positions, they were all slightly shifted to the red end of the spectrum. No matter which galaxy he looked at, all of the lines were shifted towards the red, an effect very similar to the Doppler effect, in which approaching sound waves, e.g. of a police-car siren, are squashed together and go up in frequency. When the light left the galaxies, the Universe was at a certain size, but in the intervening period of time the

Universe and indeed space itself expanded and stretched out. This stretching is seen in the spectrum as it is stretched and the absorption lines appear to move position. If space had been contracting rather than expanding, the lines would have been shifted towards the blue end of the spectrum, as we can see in a small number of galaxies in our Local Group, like the Andromeda Galaxy.

It is important to note that on a local scale within galaxy clusters the individual galaxies are often found to move towards each other, but that is because their own motion is greater than the expansion of the Universe at a local level.

The discovery of the expanding Universe is in itself an intriguing concept but it leads to an even more amazing idea. If the Universe is expanding now, then if we follow it backwards in time, at some point it must have all started from the same position and some kind of explosion occurred to drive the expansion we see today. This is the Big Bang theory and, as we saw in Chapter 1, further investigation suggests the initial explosion happened between 13 and 15 billion years ago.

All theory? Not quite, as it is possible to see evidence of the Big Bang. In 1964 two radio technicians, Arno Penzias and Robert Wilson discovered the faint echoes of the Big Bang. You can actually tune in to the signal from the Big Bang yourself if you turn on your TV and select a channel that is not tuned to a particular station so that you see the 'snow' all over the screen. The majority of the interference causing this is terrestrial but a very small proportion comes from the Big Bang.

This leaves a rather obvious yet uncomfortable question: if the Big Bang caused the Universe to expand, and given the observations from Hubble, Penzias and Wilson that looks a very strong theory, then what happened before and what actually caused the Big Bang? I once heard another scientist being asked this question; his reply was a rather cunning one. He simply explained how space and time were supposed to be part of the same thing and, while it was commonly believed space was spontaneously created at the Big Bang, it was also thought that time was created then too. So he reasoned that it was meaningless to ask what happened before since there actually was no before! This idea is not particularly new – back in the fifth century, Augustine of Hippo claimed the Universe was created 'not in time, but simultaneously with time'.

It is now a widely accepted idea that space and time are two sides to the same coin. They both form part of something called the space–time continuum. To understand how they are linked, imagine you are meeting a friend at a new bar you have found. You first need to communicate to your friend where the bar is. If the bar is on the first floor of a building you could simply specify its address and that it is on the first floor, and by doing this you are describing its position in relation to the three spatial dimensions. So far, then, you both know how to meet up 'in space' but without further information you can still very easily miss each other by being there at the wrong time. For this meeting to be a success, you also need to know about the location in the fourth dimension, time; in other words, the agreed time of your meeting. So, you see, if you only know the spatial co-ordinates (the address) you could both turn up at the wrong time, and if you only know the time coordinate (time of day) then you could turn up at completely the wrong location. One is not a lot of good without the other; they are both dimensions that describe position in the Universe.

When considering the creation of the Universe, there are implications of space and time being linked. This leads us back to that rather nasty concept that we must not really talk about – what happened before the Big Bang – because time did not exist if there was no before! If all this fanciful talk is wrong and something did actually cause the Universe to pop into existence, then what was it? There are a number of popular theories and one of them stems from a young branch of science known as string theory. In its very simplest form, string theory explains how everything in the Universe is made up of tiny strings. The strings vibrate in many different ways and the way they vibrate determines what particles we see. Like the strings on a guitar which vibrate to give us different notes, the strings in string theory vibrate in such a way as to give us the different particles which make up all the matter we see in the Universe.

String theory also suggests there may be many more dimensions than the four that we recognize (three space and one time). When this is applied to the Universe at large, the three spatial dimensions actually exist on a multi-dimensional sheet called a membrane, or 'brane' for short. It is thought that there are more than one of these branes, maybe even an infinite number, and they exist a tiny distance away from each other. The theory hints that every few trillions of years two of the branes collide with each other, sparking a new Big Bang and perhaps the creation of a new parallel Universe. It goes on to suggest that these brane collisions may have happened for an infinite length of time in the past and may well continue to happen for all eternity. Unfortunately, there is no solid evidence for any theory like this or any others that attempt to explain the cause of the Big Bang, so this question will, I fear, have to remain unanswered.

Instantly after the Big Bang, the Universe was a seething mass of energy with unimaginably high temperatures and pressures. There were none of the particles of the sort we see today and no sign of the building block of the Universe, the hydrogen atom. The four forces that dominate the world today (gravitational, electromagnetism, strong nuclear and weak nuclear) were all part of one unified force. At 10^{-43} seconds (1 million trillion trillion trillionth of a second), an incredibly small period of time after the Big Bang, gravity became separated from the other three forces. Compared to the others, gravity is quite weak but so much more dominant in our lives simply because its effect can be felt over vast areas of space.

Shortly after, another of the four forces, the strong nuclear force, became separated. This is the force that holds the particles together in the nuclei of atoms. As its name suggests, it really is a strong force, much more powerful than gravity, although the distance over which its effects can be felt is much smaller.

Prior to the separation of these two forces the Universe was a tiny fraction of a millimetre across. The separation of the strong nuclear force from the electromagnetic and weak nuclear forces effectively made the force of gravity repulsive for just 10^{-32} seconds. Do not worry about how this happened, it just did. During this 'inflationary period' the expansion rate of the Universe soared, causing a sudden increase in size by 10^{50} times from incredibly small to roughly the size of a melon. That is in comparison to its current estimated size of about 93 billion light years across. This expansion that took place in the inflationary period of the evolution of the Universe happened at staggering speed. A 'ray' of light takes 30 billionths of a second to travel 1cm in a vacuum, but during inflation the Universe expanded by 1cm every 10 billion-trillionths of a second, a lot faster.

Up until this point, the Universe has been home to temperatures and pressures so extreme we simply can't imagine what the conditions were like. After the brief but rather swift expansion, particles called quarks and electrons started to form out of the energy. Quarks and electrons are the building blocks of matter, of all the things we see in the world today; cars, houses, stars, galaxies, even you and me, all are made up of different collections and arrangements of quarks and electrons. At the same time there also arose the creation of their anti-particle equivalents: the anti-quark and the positron or anti-electron. An anti-particle is just like the normal particle but opposite in its properties. For example, the electron, one of the tiny balls whizzing around the nucleus of an atom, has a negative electric charge but its anti-matter equivalent, the positron, has a positive electric charge. Anti-quarks are just another type of anti-matter and, fortunately for us, at this stage of the development of the Universe quarks outnumbered anti-quarks by a ratio of 1,000,000,001 to 1!

You may be familiar with the fact that a meeting between matter and anti-matter causes the two to annihilate each other. As soon as the quarks and anti-quarks appear, they indeed start annihilating each other until there are just quarks left. This process floods the Universe with electro-magnetic energy, or light. But it is the remaining quarks that are most important as they will form the real building blocks of our Universe.

Things are happening quite fast now so let us move on. We can now really start to appreciate how far things have gone given that it has only been 10 billionths of a second since it all started. If you remember, there are four forces in the Universe. So far, the gravitational and strong nuclear forces are separated from the others and trying to exert their hold on proceedings. The remaining two forces – electro-magnetism (which effectively controls the attraction of negative and positive particles) and the weak nuclear force (which causes the sudden decay of one type of atom into another) – separate from each other, leaving all the forces free to shape the Universe we see today.

Just one thousandth of a second after the Big Bang the strong force finally makes its presence felt as it causes the quarks to combine to form the protons and neutrons which will finally become the central part of atoms. I've not said much about the temperature of the Universe during the early stages – frankly, it is enough to say it has been very hot. After all the changes we have just seen in the first few fractions of a second, the temperature finally starts to cool to a rather more pleasant shirt-sleeve temperature of 1 million million degrees. At these high temperatures, the Universe is a very unstable place with radiation much more dominant than matter. When we take a look at the behaviour of matter in the form of protons, neutrons or electrons at different temperatures, we find that 'cold' particles do not move around particularly fast; however, 'hot' particles are terribly hyperactive and whizz around at a rate of knots. So, if the Universe is very hot, then all the particles are extremely energetic and will not join up and bind with each other. The energy from the heat is much stronger than the energy that holds them together but, once it cools down, things start to get interesting.

Over the next few fractions of a second, the temperature decreases to around 100,000 million degrees and the conditions start to settle, eventually giving matter much more chance to get a foothold. With this change in conditions, protons and neutrons can stick together thanks to the strong nuclear force. Protons have a positive electric charge so they repel each other (remember that opposite charges attract); however, the neutrons have no charge, as their name suggests, and are said to be neutral. The Universe starts to fill with atomic nuclei composed of one proton and a number of neutrons (generally not more than three). This marks an important point in the journey: the nucleus of the hydrogen atom is formed. From around three minutes after the Big Bang, nucleosynthesis (the process that makes stars shine) starts to fuse neutrons and protons together, flooding the early Universe with a specific type of helium. This process only lasts for around eighteen minutes though, since the Universe cools to a temperature at which fusion cannot continue. At the end of this era, there are about three times as many hydrogen atoms as there are helium ones, and only small amounts of other atomic nuclei.

Now, for a much greater leap in time. Around 70,000 years after the Big Bang, tiny irregularities in the distribution of matter slowly start to get amplified and it is thought that dark matter plays a part in this process. The temperature and density of the Universe continue to drop and around 380,000 years after the Big Bang they reach a level that finally allows electrons to attach to the hydrogen and helium atoms. There is another long wait, as much as 150 million years, before the first stars start to form out of the gravitational collapse of hydrogen and helium clouds. This first generation of stars, called Population III stars, are just theoretical at the moment since none have been positively identified. It should be possible to identify them using spectroscopes, which study the light from them to reveal a lack of heavy elements and a particular quantity of hydrogen and helium in their core. This first generation of stars would have been responsible for converting the lighter elements, hydrogen and helium, into the heavy elements. A process that went on over billions of years to produce other stars, planets and even living creatures like you and me.

Some theories suggest that the Population III stars may well have existed outside the confines of the gravitational bonds of a galaxy. At the same time as their creation during the first billion years of the existence of the Universe, the collapse of larger gas clouds led to the formation of the first galaxies. One of the earliest we can see is called UDFj-39546284 and is found in the constellation Fornax in the southern hemisphere sky. It is around 13.2 billion light years away and at this great distance we are looking back in time at an object that formed just 480 million years after the Big Bang. Being able to look backwards in time like this, because of the vast distances in space, is an incredibly useful tool for us in understanding how the Universe has evolved. It is almost like an archeologist digging up layers of the ground to find out how history has changed.

Unfortunately there is a limit to how far back we can go. During the inflationary period the Universe expanded faster than the speed of light so that every part of space was rushing away from every other part of space faster than light can travel. This means if a star gave off a photon of light during the inflation phase (although no stars actually existed so soon after the Big Bang), an observer in another part of the Universe would effectively be carried away by the expansion of space at a speed faster than the light that was heading towards them. The light would never actually catch up with the observer regardless of the fact that the expansion slowed down fractions of a second later. For astronomers, the inflationary period acts as a barrier to attempts to study the early Universe as any radiation being emitted by events prior to this would never reach us here on Earth. We can only ever hope to see as far back as the inflationary period and even that may well be beyond the limits of our technology. Instead, we are left with making assumptions and intelligent guesses based on observation and possible scenarios.

As we saw in Chapter 5, the first galaxies to form, like UDFj-39546284, are thought to have been spiral in shape and, over time, gravity caused those nearest to each other to collect into clusters and superclusters. Studies of the distribution of galaxies in the Universe today still reveal the clusters and superclusters whose general structure was laid out in the Universe almost 13 billion years ago. In those same studies we can see signs of galactic collisions which caused spiral galaxies to merge, forming larger elliptical galaxies.

In the galaxies Population II stars formed and then, as they died, they seeded the galaxy with heavier elements that eventually formed the majority of the stars we see today, the Population I stars. Not only are the heavy elements found inside the stars to a lesser degree but they are also found forming huge discs around the hot young stars called proto-planetary discs, which will eventually form planets just like ours. From this point on, stars continue to evolve and die, planets come and go, galaxies merge and maybe even life will evolve in some other remote corner of the Universe. The cosmos will continue to expand at a rate that seems to be slowly getting faster and faster until, many billions of years in the future, it meets its fate.

It is amazing to think that all of this, our knowledge of

how the Universe has evolved, even its very beginnings, has come from simple study of the night sky. Observations have led to theories, and theories have been tested against new observations. The process of science and its investigative nature is subjected to its most rigorous test when applied to studies of deep space and yet, while there are some mysteries still to be solved, it is slowly unpeeling the layers and unlocking the story of how our Universe evolved.

November: Northern Hemisphere Sky

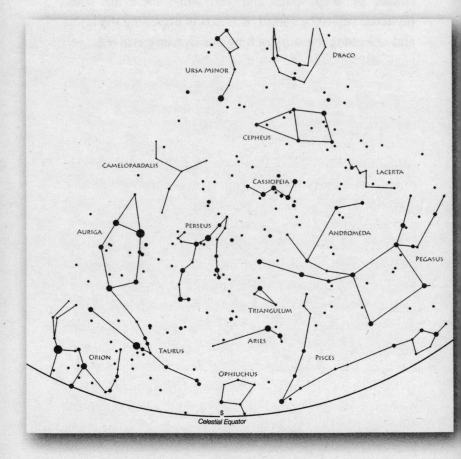

A few prominent stars dominate the November northern hemisphere sky in the east, such as the bright red Betelgeuse in Orion just north of the celestial equator, Aldebaran in Taurus to the north-west and Capella in Auriga even further north. Aldebaran along with its companion stars in the Hyades star cluster form a celestial 'V' shape which conveniently points towards another red star, called Menkar, the second-brightest star in Cetus and the starting point for our November northern sky guide.

Menkar lies just north of the celestial equator and is said to have a declination of just over 4 degrees. Menkar's distinct colour suggests it is nearing the end of its life having completed the hydrogen-burning phase and possibly even the helium phase too. It is a star which has a mass about three times that of the Sun and a diameter about 90 times greater, but eventually its outer layers will be lost to space, forming a planetary nebula, a fate that also awaits our Sun.

To the north-east of Menkar is a faint cluster of stars called the Pleiades Cluster, which looks like a tiny version of the Plough in Ursa Major. The cluster is easily visible to the naked eye and is often used as a test for eyesight and atmospheric conditions. There are at least 1000 stars in this cluster; although between six and twelve have been recorded with the naked eye by observers, binoculars will reveal up to a hundred stars and telescopes many more. It measures 15 light years across, is 440 light years from the Earth and, at this distance, appears in our sky to be about four times the diameter of the full moon. On moonless nights, it is possible to detect a hint of fuzziness around some of the brighter stars, especially Merope, the southern-most of these. Studies of the direction of travel of the cluster and its nebulosity show that the two are unrelated, that the cluster is simply moving through a dusty part of space and it is the light from the stars which is reflecting off the gas and dust molecules, much like car headlights lighting up fog.

Due west from the Pleiades Cluster is a modest 2nd magnitude orange star called Hamal, the brightest star in the constellation of Aries. About 2000 years ago one of the two points where the path of the Sun crosses the celestial equator was in Aries but it has now moved to Pisces as we saw on page 54. The rest of Aries looks like a curved line with Beta and Gamma Arietis to Hamal's west pointing south to the celestial equator. To the north-east of Hamal is the most westerly star in Aries, known as 41 Arietis and it points to the most southerly stars in Perseus.

Perseus represents another hunter, or more accurately a Greek hero sent to slay the snake-haired monster Medusa. The constellation depicts the triumphant hero grasping Medusa's severed head. The most southerly star in Perseus, which Hamal seems to point towards, is Zeta Persei and it marks one of his feet, the other being to its north-east without any prominent stars. To the north-west of Zeta is a star by the name of Algol, which is perhaps the best-known of the eclipsing binary stars. These binary star systems orbit each other with the orbital plane along our line of sight, which means we see them alternately passing in front of each other, making the combined amount of light momentarily dip. In the case of Algol there are three stars in the system but only the two brighter ones eclipse each other every 2 days, 20 hours and 49 minutes, when the star system dips by just over 1 magnitude in brightness from 2.1 to 3.4. Moving north from Algol is a line formed by Misam and Iota Persei. North of Zeta is a rather more jagged line of stars made up of Menkib, Adid Australis, Nu Persei and Adid Borealis.

Between Menkib and Adid Australis is a large emission nebula known as NGC1499, or the California Nebula. Emission nebulae, as their name suggests, are visible because they emit their own light with a characteristic red glow that is the result of energized hydrogen atoms. This fine example in Perseus gets its name from a striking similarity to the shape of the state of California but is somewhat longer, measuring 100 light years. Even though it is a 5th magnitude object, it is visually challenging to observe without optical aids; in fact, exceptionally dark and clear skies are needed to see it. Using a wide-field telescope and an h-beta (hydrogen-beta) filter will help significantly.

To the north of Algol and Zeta Persei is the brightest star in the constellation, called Mirfak, or Alpha Persei. It is also the brightest star in a cluster of stars known as the Alpha Persei Moving Group, which is a collection of about a hundred stars that all broadly share the same speed and direction of movement. They lie at a distance of about 600 light years and cover a volume of space measuring just over 30 light years in diameter. The other stars in the cluster can be seen scattered around Mirfak from a nice dark site.

Mirfak is also at the centre of a curved line of stars that points to the north-west and towards the constellation of Cassiopeia, which is seen as a giant 'W' in the sky (or 'M' from low southern hemisphere latitudes just skirting the horizon). Take a glance at the area of sky between the two and you will see what seems to be a couple of hazy patches. This is the famous Perseus Double Cluster of stars and from a dark site they can easily be seen with the naked eye, with each cluster covering an area of sky about the same size as the full moon. The clusters, which have the catalogue numbers NGC869 and NGC884, are just a couple of hundred light years apart and nearly 7000 light years away from us. Spectral studies show that their light is shifted towards the blue end of the spectrum so they are heading towards us at a speed of roughly 20km per second. Open clusters like these are usually young but these two are among the youngest, with an average age between them of just 4.3 million years. Through even the smallest of telescopes with a low magnification, this pair of clusters is visible in the same field of view and looks mesmerizing. Even modest binoculars offer an impressive sight.

Returning to Perseus and looking to its north-east is the large, faint constellation called Camelopardalis, whose most westerly stars form a line which points to Perseus in the south and Polaris, the North Pole Star, in the north. This line is not the most prominent, however, and is marked at its most southerly point by the variable star CS Camelopardalis, which appears as a faint yellow-white star. To the south-west of this star is Miram, an orange star which sits at the northern tip of Perseus and is a stunning binary system with contrasting gold and blue stars. Between CS Camelopardalis and Miram is the radiant, or the point in the sky, that the Perseid meteors all seem to come from during the annual meteor shower seen each August.

November: Southern Hemisphere Sky

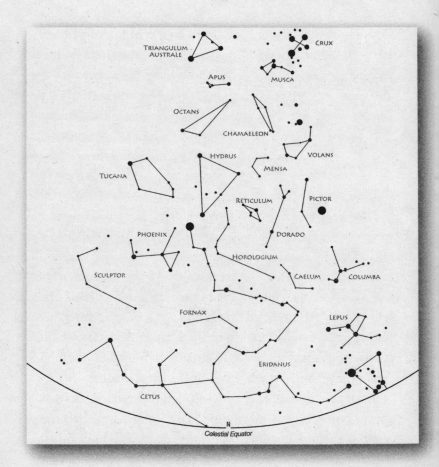

Sitting on the celestial equator and therefore with a declination of zero degrees is M77 in the constellation of Cetus. Visually it appears as a 9th magnitude spiral galaxy, although it is officially classed as a barred spiral. It is easy to find just a degree to the south-west of Delta Ceti and about 10 degrees to the south-east of the more prominent Alpha Ceti. Through a small telescope with an aperture of perhaps 10cm the central nucleus appears to dominate the view but with larger telescopes the three prominent spiral arms come into view. Higher magnification reveals an almost pinpoint core to the galaxy, which hints at something a little different. The core of M77 is what is known as an active galactic nucleus which is a nucleus that emits an incredible amount of energy. It is thought that a supermassive black hole may sit at its centre, driving the emissions.

Further south of M77 is a rather bland portion of the sky which is filled only by the faint stars of the constellation of Eridanus, the river, as it meanders around the sky. It starts in the west just to the north-east of the strikingly bright star Rigel at the foot of Orion, the hunter. The star marking the start of the river is called Cursa and is the second-brightest star in the constellation. There is then a line of stars which curves towards the celestial equator, ending at Gamma Eridani, the third-brightest star in the constellation.

To the north-east of Gamma Eridani by about 5 degrees are a couple of 4th magnitude stars lying parallel to the celestial equator and about 3 degrees apart. Epsilon Eridani is the most easterly of the two stars, is fractionally fainter and at 10.5 light years away is the tenth-closest star to our own. A dust disc has been discovered around this star stretching to a distance of around forty astronomical units but it seems to have been depleted closer in, which shows it has already condensed to form planets. One of the planets has been discovered to be about 1.5 times the mass of Jupiter and orbiting at an average distance of 553 million kilometres, compared to the Earth's orbital radius of 150 million kilometres.

From Gamma Eridani, the curving lines of Eridanus sweep towards the west before running back to the east towards Achernar, the brightest star in the constellation. To the south of Epsilon Eridani by about 20 degrees and of almost identical brightness is a star called Dalim, the brightest star in the constellation of Fornax. Also known as Alpha Fornacis, this star is a binary star system only 46 light years away and composed of two yellow stars, both of which are visible with a modest pair of binoculars. Less than a degree to its north is the location of UDFj-39546284, the oldest identified object in the Universe. Unfortunately it is not possible to see it with any amateur equipment; in fact even the Hubble Space Telescope only reveals it as a faint red smudge. Fornax is a small constellation that resembles a shallow triangle with the star at its tip, Beta Fornacis, pointing towards the South Celestial Pole. The eastern point of the triangle is marked by Nu Fornacis.

One little treat in the centre of the triangle of Fornax is a fabulous barred spiral galaxy called NGC1097. Like M77 in Cetus at the start of this guide, NGC1097 is another active galaxy but of a slightly different type. It lies about 45 million light years away and, from our location, we are looking almost straight-on to it. This means its bar and spiral arms are nicely visible, but at 9th magnitude it will just be seen as a fuzzy disc with a small telescope. Larger telescopes of 20cm or more are needed to detect its arms and central bar protruding from the nucleus.

Fornax A, the other main galaxy in Fornax, is found around 8 degrees due south of Dalim. Unlike NGC1097, this is an elliptical galaxy, shaped like a rugby ball or an egg. Fornax A is actually the name given to the source of strong radio emissions from the parent galaxy, NGC1316, not from the galaxy itself. It lies at a distance of 75 million light years away, about 30 million light years further than NGC1097 to its north. Long-exposure images from telescopes like the Hubble Space Telescope reveal dust lanes that are unusual in elliptical galaxies and are generally only found in spiral or barred spiral galaxies. It is thought that the dust lanes are the remnants of a spiral galaxy that collided with NGC1316 over 100 million years ago. The galaxy is easily visible with small telescopes but the dust lanes are rarely seen through amateur instruments.

Eridanus curves around to the south of Fornax again, and to the south-east of NGC1316 is Theta Eridani, a 3rd magnitude star which is home to a beautiful binary star that is ideal for small telescopes. At a distance of 160 light years, the two stars appear to be separated in the sky by a tiny distance equal to 1/450th the width of a finger at arm's length and observation requires a telescope.

To the south of meandering Eridanus, which dominates

the sky to the south of the celestial equator, is a small, faint constellation called Horologium. It is supposed to represent a pendulum clock and is one of a number of southern hemisphere constellations which symbolize scientific equipment, along with Telescopium and Microscopium. Alpha Horologii is its brightest star, at 4th magnitude, and is found just under 10 degrees to the west of Acamar in Eridanus. It appears orange in the sky, which reveals that it is a moderately cool star with a temperature of around 4600 degrees. The rest of the stars in the constellation are around 5th magnitude, so fainter, and lie to the south-east of Alpha Horologii.

Roughly halfway between Alpha Horologii and Achernar, and about 20 degrees to the south-east, is NGC1261, a fine example of a globular cluster. This 8th magnitude cluster is visible as a fuzzy star with binoculars and some of the outermost stars can be resolved individually in 15cm telescopes. To see the whole cluster and resolve stars in the core requires a telescope with an aperture of at least 25cm. In these larger instruments it appears as a very condensed cluster in among a rich field of foreground stars.

To the south-west of Horologium is what seems to be a diffuse patch of light almost like a breakaway portion of the Milky Way: it is one of our satellite galaxies, the Large Magellanic Cloud (LMC), and is very similar in nature to the Small Magellanic Cloud to the south-east. Halfway between Horologium and the LMC is a small constellation called Reticulum, which looks like a squashed square, or parallelogram, made up of 3rd and 4th magnitude stars. Alpha

Reticuli is the corner star nearest to the LMC and is noticeably brighter than the rest. About 10 degrees along the line between Alpha Reticuli and Achernar in Eridanus is a beautiful double star called Zeta Reticuli, visible to the naked eye. It lies 39 light years away and the separation between the two Sun-like yellow stars is around 9000 times the Earth–Sun distance.

TWELVE

The End of the World

W<small>E ARE ALL</small> used to the idea that time moves on, the seasons come and go and even we humans do not live for ever. Yet for some reason, most people seem to assume that the Earth and the Universe itself will last for all time. Alas, our home, like the Universe it lives in, has a finite life and an end will come to everything, in one way or another. The nature of that ending has been the subject of many a scientific debate yet we still do not know what form the end of the world will take. When considering this question, which is perhaps one of the biggest questions in science, it is appropriate to consider not just the end of the Earth but also the end of the Universe. Having an understanding of what might bring about the end of things on our home planet leads us to a better understanding of stellar evolution and ultimately to the fate of the Universe.

Before 1992, the only planets we knew of in the Universe were those that orbited the Sun in our Solar System. Given that there are more stars in the Universe than grains of sand on Earth, it seems likely that at least one of them would have a family of planets like our Sun. Over 700 planets have been discovered around other stars now and it seems that the formation of planets is fairly commonplace after all. By observing the vast range of stars at different stages of their lives, we can see that the death of stars is also a common occurrence. Such an event would herald disaster for any planet in orbit around it and, indeed, when our Sun dies it will spell the end for life here on Earth, and possibly even the end of the Earth itself.

All stars have a zone around them of varying distance, dependent on how hot the star is, which has been called the 'Habitable Zone' or the 'Goldilocks Zone' because it is at this distance that conditions are suitable for life, not too hot and not too cold. In this region the temperature is just right for water, or more precisely H_2O (referring to two hydrogen atoms and one of oxygen), to exist in its three states: in the solid form as ice, the liquid form as water and in its gaseous state in the atmosphere, and it is this which makes the Earth ideal for life. If the Earth moved out of the Goldilocks Zone or if the zone itself shifted for some reason, then our planet would become uninhabitable and the complex life forms that exist here, including us, would very likely come to an end.

This may all sound like a far-off distant worry, but if the Earth is going to die at some point in the future, then for humanity to survive will take an incredible amount of planning, and understanding the problem is the first step in this process, even though it may be billions of years away.

The reality is that the fate of the Earth as a hospitable planet rests solely on the evolutionary path of the Sun, and while at the moment this is stable, it will not be so for ever.

The current stability of the Sun comes from the process in its core fusing hydrogen into helium, as we have already seen. An outward force called thermonuclear pressure from the nuclear reaction tries to push the Sun apart but the balancing force of gravity tries to collapse it. The net result is that the Sun stays in its current state, but there will come a time when it runs out of hydrogen in its core and starts fusing helium atoms. The change of reaction will cause an increase in the thermonuclear pressure which momentarily wins over gravity and the Sun will swell in size. As the Sun goes through this red giant phase it is expected to increase in size by 250 times its present radius, but that means, as it swells, it will swallow up Mercury, Venus and possibly even the Earth. This all sounds quite fanciful but it is thanks to a technique called spectroscopy that we can study the chemical composition of the Sun from looking at its light. We can then compare what we see with other stars in the night sky and estimate its age and roughly how long it will be before it runs out of fuel and, when it does, what is likely to happen to it.

The mechanics of the process are actually quite complicated since initially the Earth will move to a more distant orbit because the Sun will have lost mass as it expands and have less gravitational pull. It is possible that planets may get ejected from the Solar System when this happens but it is very unlikely to happen in our case. Instead, the tug from the Earth on the Sun will cause a tidal bulge to appear on the Sun, which will lag slightly behind the line between the two objects. The bulge will drag on the Earth, slowing its orbital speed and causing it to slowly fall back closer to the Sun. This is of no immediate concern as it is thought this will not happen for at least 5 billion years, maybe even up to 7 billion years, depending on how much hydrogen remains in the Sun's core. A great example of a star going through this process is the red giant Betelgeuse as we saw in more detail in Chapter 9, which lies in the constellation of Orion. It is a star nearing the end of its life that has increased in size so much that if it were at the centre of our Solar System it would have swallowed up the Earth.

There is, however, a more pressing concern for the Earth, although it is still thought to be a billion years away. The Sun is considered to be stable currently but it is slowly increasing in luminosity and temperature, by about 10 per cent every billion years. It is likely that in another billion years the Earth will be too hot for liquid water to exist, and if liquid water does not exist then entire ecosystems will break down and it will be much more difficult for life to thrive or even survive.

It is quite probable that it will be the evolution of the Sun that ultimately brings about the demise of life on Earth and one day in the future completely destroys it, but it is not just the Sun we should be keeping an eye on as there is more of an imminent threat from silent, dark and potentially deadly asteroids. You only have to look at objects in the Solar System – the Moon, Mercury, Mars and even the Earth – to see the effects of pieces of rock flying around unchecked. The Moon has countless examples of tiny impacts where small rocks have crashed into it, leaving dents that we see as the craters pockmarking its surface. By studying the distribution and formation of the craters on the Moon we can see that impacts happened with more frequency during the Solar System's early history. The Earth is moderately protected from such events by its atmosphere, which tends to destroy the rocks as they fall to Earth as meteors or shooting stars. We are generally safe from smaller impacts, but what about larger ones? Those are the ones we need to keep a lookout for.

We can see evidence of large impact events, where much larger rocks tear through the atmosphere and make it to the surface, throughout Earth's history. The most famous of these are the asteroids that crashed to Earth causing the extinction of the dinosaurs, or the asteroid around 60m in diameter that exploded about five kilometres over Tunguska, Russia, in 1908, flattening trees and breaking windows for hundreds of kilometres around. Over time it is possible that we will be getting safer, as many of the larger rocks should have hit by now but, even if they have, there may still be others which could pose a threat.

There are around 9000 near-Earth asteroids, as they are known, with orbits which take them close to or, in some cases, across the orbit of Earth. These orbits are very well understood and there are none that pose a significant danger in at least the next few hundred years. It is not the ones we know about that should worry us though, but the ones we have not discovered yet, and there may be plenty out there. The asteroid belt between Mars and Jupiter occasionally ejects objects through collisions between asteroids but the greater concern is the Kuiper Belt, beyond the orbit of Neptune, where there are thousands of pieces several hundred kilometres across. Generally they are dark and difficult to spot and it is left to a mere handful of automatic sky surveys such as LINEAR (Lincoln Near Earth Asteroid Research) and NEAT (Near Earth Asteroid Tracking) and whole armies of amateur astronomers around the world to scour the skies. The task is daunting, however, and is rather like looking for a needle in a haystack at night, but fortunately other automated searches are being developed to identify risks early enough to allow some sort of defensive action to be taken.

In reality the asteroid threat is unlikely to destroy the Earth completely but there is a very good chance that a large-scale impact could erase life from its surface. Impacts of this type thankfully do not come along that often but, when they do, they cause worldwide devastation. The last mass extinction caused by asteroid impact was 65.5 million years ago and it brought with it the demise of the dinosaurs and 75 per cent of all species, but on average the Earth sees impacts from asteroids at least 5km in diameter every 10 million years. With impacts of this scale, millions of kilograms of debris can be thrown into the atmosphere, blocking light from reaching the surface and plunging the Earth into a global winter that might last hundreds if not thousands of years. Tsunamis, earthquakes and extreme geological activity are just some of the effects of such large-scale impacts and it could take several million years for life to recover.

There is another risk to the sustainability of life on Earth and it is thought to come either from the collision of two collapsed stars or from the collapse of a supermassive, fast-rotating star at the end of its life. This event is seen as a brief burst of gamma radiation which can last from fractions of a second to a few minutes. It is believed that a highly focused beam shoots out into space, and if you are in the line of sight, you see the burst. Following the initial blast, known as a gamma ray burster, there is an afterglow which can last several hours longer and is seen in different wavelengths. The process is still not fully understood but these events are observed roughly once a day and are thought to emit 10 quadrillion times more energy than the Sun, making them the most luminous and violent events ever observed.

The bursters were first discovered in the 1960s by satellites that were designed to detect gamma radiation bursts from nuclear weapons testing. With the detection of a burst in July 1967 more accurate satellites were launched to investigate further, and an additional fifteen GRBs were picked up. By analysing the timings and positions it was possible to find an approximate location which ruled them out as man-made in origin. The closest burster, GRB031203, was detected in December 2003 in the constellation of Puppis at a distance of 1.3 billion light years, but to date none have been observed in our galaxy. Fortunately, at those distances they are nothing more than a scientific curiosity, but if one occurs closer to home then we might be in trouble.

Most of the stars that can be seen with the naked eye are no more than just a few thousand light years away, so if a gamma ray burster went off among the stars it might be bad news for life on Earth. We can tell if a star is near the end of its life and a candidate for such an event, but we may be unable to detect a binary star system on the brink. If one did go off within a couple of thousand light years from us and send the burst of radiation in our direction then it is likely that it would outshine the Sun for a brief moment and the initial burst would fry the atmosphere, creating elements that would destroy the ozone layer. With no ozone layer the full force of ultra-violet radiation from the Sun would make its way to the surface of our home, cause severe cases of skin cancer, kill off the plankton at the bottom of the food chain and destroy oxygen-creating plants. The result would be catastrophic.

Whether the end of the world is considered to be Earth's complete destruction or it being left in an uninhabitable state, there is no dispute that the ultimate end must be finally and irreversibly defined by the end of the Universe itself. As we have seen, the discovery by Edwin Hubble that the Universe is expanding led to the conclusion that it formed in the Big Bang, but its fate is a little harder to pin down. It may continue to expand for ever or eventually stop expanding and collapse back in on itself. The answer will be found by identifying how much matter is in the Universe, or more accurately its density. Density is a term that refers to the amount of matter (mass) per unit volume and measuring the density of the Universe has proven to be quite tricky. Of course, it is impossible to actually measure the density of the entire Universe so, instead, calculations of the volume and mass of a small portion of space are made and used to extrapolate the overall density.

 If we measure the speed of the expansion of the Universe it is possible to determine what the density of the Universe needs to be in order for there to be enough matter and hence gravity to arrest the expansion. We can visualize how matter interacts with the Universe by way of gravity by imagining a sheet of rubber. If a large heavy object were placed on the sheet it would sink into it making a depression. In this analogy, the heavy object could represent a star and the sheet, the fabric of space. The depression would represent gravity and any object travelling through it or close by would get drawn towards the massive object.

The critical density is a term used to define how much material is needed for there to be enough gravity to halt the expansion, and its value is 1 million trillion trillionth of a gram per centimetre. This is an incredibly small amount of material but it is the average density that is important and, of course, in some regions of space, such as in the centre of a star, the density is significantly higher. If it turns out that the actual average density is less than the critical density then the force of gravity will not be enough to halt the expansion and the Universe will continue to expand for all eternity. It may be that the density is more than the critical density, in which case gravity will ultimately overcome the expansion, stop it and a period of contraction will follow. There is a third option where the density of the Universe is exactly the same as the critical density, in which case the Universe is considered 'flat' and it will continue expanding, but at a slowly decelerating rate, yet will never actually stop.

If all of the visible matter in the Universe is added up it seems to fall far short of the quantity needed to halt the expansion, yet some theories suggest there should be just enough matter to produce a flat Universe that is locked in an ever-decreasing rate of expansion. One possible explanation for this lies with a rather strange and exotic material mentioned in Chapter 10, called dark matter, which was first discussed by Jan Oort in 1932. His theory, which suggested there was a new type of material that would be hard to directly detect, was in response to the observed orbital speed of stars in the Milky Way and other galaxies, which was higher than expected – in fact, the stars should have been flung out of the galaxy at the speed they were travelling yet they clearly were not. As we touched upon briefly in February's sky guide, gravitational lenses are another phenomenon which hints at the presence of more material than we are able to detect. These natural lenses are visible to us when a massive galaxy or galaxy cluster distorts the light from a more distant object, creating a lensed image of it. By measuring the amount of distortion, it is possible to calculate the mass of the object doing the lensing and observations have shown that there is significantly more matter than we can see directly. The exact nature of dark matter is still not well understood but it is now thought to make up 83 per cent of the matter in the Universe.

There is one further twist in the search for the end of the Universe and it relates back to observations in 1988 of supernovae in distant galaxies. Specifically, type 1a supernovae are thought to result from the violent explosion of a white dwarf star in a binary star system and, due to the nature of the event, will always give off the same amount of energy and shine with a very specific brightness. Comparing the apparent brightness in the sky with the actual brightness allows the distance to be calculated. Another technique for measuring vast distances in space relies upon studying the spectrum of the object and calculating the apparent shift in the position of characteristic lines in the spectrum. Doing this allows us to determine the speed away from us and with a little mathematics we can determine the distance of the object, as we saw in Chapter 5. When this was done for the supernovae, there was a problem: they were found to be 15 per cent further away than they should have been. There are a few explanations for this, all but one of which have now been discounted: that the rate of expansion of the Universe is accelerating.

The acceleration is explained by a concept called dark energy, which exerts a pushing force on the Universe in competition to the pulling force of gravity. There are two key theories to explain dark energy, one of which has its origins in the very early stages of the evolution of the Universe. A few fractions of a second after the Big Bang there was a brief period when gravity effectively became repulsive causing the

Universe to expand faster than the speed of light. It is thought that dark energy may be the aftermath of this 'inflationary' period.

Another possible explanation is that the apparently empty 'space' that fills the gaps between the galaxies is far from empty and, instead, is full of new, temporary particles that constantly appear and disappear. When this theory is investigated it seems there is far too much dark energy than is required to drive the expansion we see in the Universe. The alternative to these two theories is simply that our theory of gravity is plain wrong. An uncomfortable statement to make, but science is as much about proving things to be wrong as it is proving them to be right, so maybe, just maybe, we need to look at the theory of gravity in more detail. Taking into account the observations of the expansion of the Universe and the estimates of matter that we can see in the sky, it seems that dark energy must represent 70 per cent of the Universe with dark matter making up 25 per cent and what we amusingly class as 'normal matter' a mere 5 per cent.

Whether the Universe continues expanding or not will determine its ultimate fate and looking at the latest observations it looks like it will expand indefinitely. If it does continue to expand for ever it could face a fate known as 'the Big Freeze'. As it continues to expand without end, it would cool, ultimately reaching a temperature that is too cold for life to exist. But it gets more gloomy than that. Slowly, over billions of years, the formation of new stars would cease, and the remaining stars would run out of fuel and, one by one, would die. The Universe would get darker and darker as fewer and fewer stars were present to light its distant corners. They might all eventually end up inside black holes, which themselves might slowly radiate energy off into space until they too dissolved and the Universe reached a cold, uniform temperature, dark and bleak. If this is to be the fate of the Universe, then it is thought this state will be encountered in around 1 million billion years (the current age of the Universe is 13.7 billion years).

It is quite easy, yet perhaps saddening, to think of the Universe ending in this way, but an alternative theory comes from research into dark matter and quantum mechanics. It suggests the Universe might be oscillating, which means that the current period of expansion may eventually be halted until reversed into a period of contraction. The Universe would shrink down to a point just before it becomes a singularity, a point with zero volume and infinite density where gravity may briefly become repulsive enough to kick-start another Big Bang. The idea of the oscillating Universe goes quite some way to resolving the unanswered question of what happened before the Big Bang, yet it does leave the rather unpalatable possibility that we could be held in an infinite loop of Big Bang followed by expansion, contraction and then a big crunch before the cycle continues.

The idea of an oscillating Universe is a step on from the idea of a big crunch, where the Universe collapses back into a singularity, ending at this point. This is not a popular idea among many scientists as it leaves the awkward questions of what happened before and what happens after. There are other equally unpopular theories to explain the fate of the Universe, such as the big rip, whereby the rate of expansion accelerates further until all matter, large or small, is ripped apart into individual particles; or the multiverse theory, in which an infinite number of Universes exist, popping in and out of existence as they are all formed in a big bang and destroyed in a big crunch.

Regardless of how the Universe will end, it seems our home, the planet Earth, is ultimately doomed. In billions of years' time, the Sun will die, expanding to colossal proportions as a red giant and possibly swallowing up the Earth. Even if the Earth survives the fiery death of the Sun it will join the rest of the matter in the Universe in a common fate. Exactly what form that fate will take is, as yet, a matter for conjecture, but the answer will be found in a deeper understanding of the nature of matter and energy. One of Einstein's famous equations explains that energy and matter are interchangeable and it seems certain that the Universe started, and will end, as energy. For a period of time, that energy was transformed into matter through some beautiful and quite elegant processes, but understanding the fate of the matter we see in the night sky today will undoubtedly keep scientists busy for centuries to come.

December: Northern Hemisphere Sky

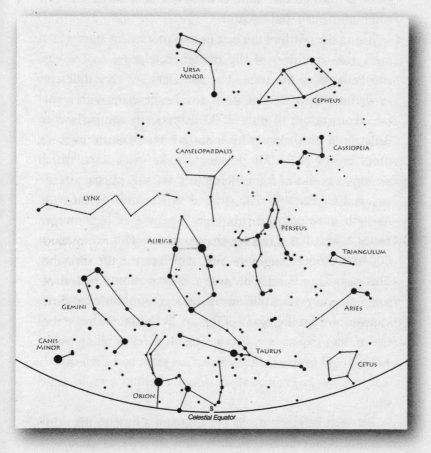

The northern and southern hemisphere skies share a very well-known constellation this month, Orion. The great hunter straddles the celestial equator with his shoulders and head in the northern hemisphere sky and his belt and feet in the southern hemisphere. The best place to start this star guide is from Orion's famous three-star belt, which lies just south of the celestial equator.

Just to the north of the belt lie two prominent stars which mark the shoulders of the hunter, Betelgeuse to the east and Bellatrix to the west. The two stars are very different: as we have seen, Betelgeuse is a red supergiant with a surface temperature of only 3500 degrees, in comparison to Bellatrix, a blue giant with a surface temperature over six times higher at 21,500 degrees. Unlike most stars, which we see as points of light, Betelgeuse actually appears in the sky as a measurable disc, second in size only to the Sun. As with most stars of this type, Betelgeuse's light output varies, and it has a regular change in diameter, from about 500 to 900 times that of the Sun, which causes the variation. This instability is seen because it is an ageing star, leading some astronomers to believe it may go supernova and die violently within the next 1000 years. Bellatrix, to the west of Orion, can expect a different fate in a few million years, when it will evolve into a smaller red giant and then quietly lose its outer layers into space, leaving behind a white dwarf star.

Between Betelgeuse and Bellatrix and a little to the north is a tiny, faint grouping of three stars with the brighter one, by the name of Lambda Orionis, marking the hunter's head.

The remainder of the northern part of the constellation depicts a club being held aloft from Betelgeuse, marked by Mu Orionis to the north-east, and then two pairs of stars, each of which is roughly parallel with the celestial equator and heads to the north from Mu Orionis. A shield is then depicted by a curve of faint stars centred on Pi Orionis to the west of Bellatrix.

Another prominent red star, Aldebaran, can be seen to the north-west of Orion and is the brightest star in the constellation of Taurus. A great way to confirm that you have found Aldebaran is to follow a line extending north-west from the three-star belt of Orion, which points straight to it. Its colour, which is similar to that of Betelgeuse, tells us that it is a star nearing the end of its life, and at a distance of 65 light years it is likely to be the best candidate for observing what may be the fate of our Sun.

As we have seen, Aldebaran sits in a prominent 'V' shape of stars known as the Hyades Cluster, though the cluster is not related in any way to Aldebaran, which is almost 100 light years closer to us. It is a group of around 400 stars that have been shown to share the same motion through space and have formed at the same time out of the same molecular cloud. At 625 million years old, the Hyades Cluster is one of the oldest known open clusters since most tend to disperse before 50 million years. Only those clusters a good distance from the gravitational disturbance of the galactic centre appear to survive for longer, although it does seem to be losing the odd star, such as one now in the constellation of Horologium in the southern hemisphere sky.

Extending the southern half of the 'V' of the head of Taurus towards the east leads to Zeta Tauri. Elnath, the brighter star to its north, and Zeta Tauri together represent the horns of the bull. Zeta Tauri is a hot blue giant star with a temperature of 22,000 degrees and spins at such a rate that it is losing matter to the surrounding space, forming an accretion disc. Elnath is cooler at just 13,500 degrees.

Zeta Tauri is a particularly well-known star because it acts as a useful guide for finding the Crab Nebula, or M1, which is the remnant of a star that exploded as a supernova in 1054. It is easily found after locating Zeta Tauri because it is just 1 degree to the north-west. When the original star exploded as a supernova it gave off so much light that it was visible to the naked eye during daylight. The light has now faded to a much fainter 8th magnitude, which means small telescopes are needed to pick it up. From dark sites on a moonless night it is just possible to see it as a misty patch with a pair of 10x50 binoculars but a telescope is definitely the best instrument for this object, a 15cm aperture giving a reasonable view. Even then, it will only look like a slightly misshapen oval with a hint of dark and light patches. The nebulosity is all that is left now of the outer layers of the star which exploded, although deep in the core of the nebula is a pulsar, a stellar corpse spinning thirty times every second.

Elnath, the star marking the northern tip of the horn, is on the border of the constellation of Auriga further north. Auriga appears in the sky as a slightly squashed pentagon and lies in a dark portion of the Milky Way. Directly north of Elnath is Capella, the brightest star in the constellation and a well-known binary star system. The two stars are similar to the Sun and have comparable surface temperatures, although both are about ten times its size. They lie about 96 million kilometres apart and at a distance of 42 light years amateur telescopes will not separate them visually.

To the south of Capella is a tiny triangle of three stars known as the Kids and marked by Epsilon Aurigae at the northern tip and Eta and Zeta Aurigae at the triangle's base. Epsilon Aurigae is a peculiar eclipsing binary star whose main star is a yellow supergiant about twenty times the mass of the Sun. As an eclipsing binary we can tell it is orbited by another star, or, more accurately, that together they orbit a common centre of gravity. These eclipses are strange because they happen every 27.1 years and last for almost two years. Current theory suggests that the companion is surrounded by a dust ring which is also eclipsing the main star. This means the eclipses last much longer than they would normally do if just the star was blocking the light.

South of the Kids grouping is Al Kab, the third-brightest star in the constellation of Auriga, to the east of Capella is Menkalinan and to its south is Theta Aurigae. Within the lines formed by these stars are two open star clusters, M36 and M38, which we saw briefly in January's sky guide. M38 can be located on a line between Theta Aurigae and Al Kab and at 6th magnitude is visible to the naked eye when viewed from a dark site. There are about a hundred or so stars in this cluster, many of which can be seen through a modest telescope against the background of stars from the

Milky Way. It lies at a distance of 4200 light years, making it a little more distant than its companion, M36.

Also known as the Pinwheel Cluster, M36 can be found around 2 degrees to the south-east of M38 and is fractionally brighter. It is also a little smaller than M38 and contains only about sixty stars, all of which seem to be hot young stars compared to M38, which surprisingly for an open cluster has its share of ageing red giant stars. One other cluster is prominent in Auriga, M37, and it is considered by many to be the best open cluster in the sky. It has around 500 member stars, covering an area about the size of the full moon, and of the three clusters in Auriga is the most distant, at 4600 light years. It is also the oldest of the three clusters, at an age of 300 million years, and like M38 some of its stars have already evolved into red giants. M37 is found slightly to the east of the line between Theta Aurigae and Elnath and about halfway along.

December: Southern Hemisphere Sky

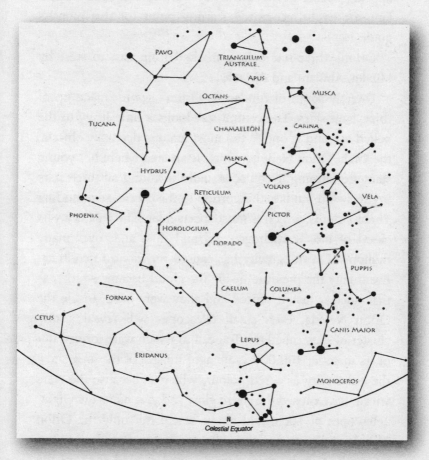

Orion is prominently placed in the December sky and is found right on the celestial equator with his head and shoulders to the north and famous belt and legs to the south. For southern hemisphere observers, the hunter appears upside down in the sky but the belt stars make a fantastic easily recognized starting point for this month's guide.

Orion's three-star belt is marked from east to west by Mitaka, Alnilam and Alnitak.

Directly south of Alnilam is Orion's sword, made up of three faint stars. The central star looks a little fuzzy to the naked eye and is one of the most famous deep-sky objects, the Great Orion Nebula, which is an area where hot young stars are forming. Nebulae actually surround all three stars in the sword but it is the portion in the centre, surrounding Theta Orionis, which is most spectacular. The area is a vast cloud of mostly hydrogen gas and dust and, over many millions of years, gravity has caused the cloud to contract. Eventually the pressure inside the cloud became so intense that nuclear fusion started and stars were born. Inside the Orion Nebula, even small telescopes will reveal a tiny cluster of stars called the Trapezium which were formed out of its material 100,000 years ago. Images of the area show the nebulosity as a red colour, which is because the stars within have caused the gas to glow and give off its own light. Telescopes of 10cm and above reveal not only the Orion Nebula and its many dark rifts and knots, but also M43 to its north. Nebulosity can be detected around some of the other stars in the sword and will appear blue in images.

These are reflection nebulae, so called because they shine by reflecting starlight rather than emitting their own.

To the north-east of the sword lies the eastern foot of Orion, marked by the bright white supergiant star Rigel. It is thought that Rigel is one of the most luminous stars in this region of the Milky Way and it shines with a brightness equivalent to around 50,000 Suns. Like many high-mass stars, Rigel has evolved quickly and has already started to run out of hydrogen in its core so is now fusing helium into carbon. Marking the western foot is the sixth-brightest star in Orion, Saiph, a hot blue-white supergiant a little over twice as hot as Rigel at around 26,000 degrees.

South of Orion is a relatively faint constellation known as Lepus with its two brightest stars, Arneb and Nihal, which can both be seen to the south-east of Saiph and form a line that points broadly to the south, almost at right angles to the celestial equator. Of the two, Arneb is the pale yellow star to the north and nearest to Orion, and Nihal is the more orange-coloured star further south.

A rather inconspicuous jewel of the sky lies on the eastern border of Lepus and is found by drawing a line between Arneb and Mu Leporis (the star to the east of Arneb by a little over 5 degrees). Extend the line beyond Mu Leporis by about 10 degrees and at this location is a beautiful red star, named Hind's Crimson Star, which is such a deep red colour that it has often been said to resemble a drop of blood against the blackness of space. Even Betelgeuse in Orion or Aldebaran in Taurus are no competition. This star, also known as R Leporis, is a variable star which takes about 430 days to slowly change its brightness from its faintest at magnitude 11.7, requiring a telescope to detect it, to its brightest at 5.5, making it just visible to the naked eye from a dark site and an easy target with binoculars. Its maximum brightness also seems to vary over a period of about forty years, from 5.5 to 6.5. The intensity of the red colour appears to increase as the star becomes fainter through its 430-day cycle.

The reason for the variability is that R Leporis is a carbon star, i.e. it brings carbon to its outer atmosphere through convection currents that build up to a thick carbon envelope through which the starlight must travel. The nature of carbon and its interaction with light means that it absorbs blue light and allows red light through so the star appears particularly red. Every so often, usually at intervals of just over a year, the carbon build-up is ejected from the star by some means, restoring the star to its usual visibility of around 6th magnitude. It lies at a distance of about 1100 light years and is thought to be 250 times the diameter of the Sun. Eventually its outer layers will escape into space, turning the star into a planetary nebula and seeding the galaxy with heavy elements that will ultimately lead to a new generation of stars. It is from stars like R Leporis, which have a high carbon content, that star systems with rocky planets form and provide the basis for life to evolve.

Using Arneb as a starting point again, extend a line south-wards through Nihal and about the same distance on is M79, a very nice 8th magnitude globular cluster. It is best viewed through at least a 20cm telescope so that many of the

stars in the cluster can be seen individually. Anything
smaller than this will struggle to resolve the stars and so will
show no more than just a mottled patch of light. M79 is in
an area of sky that does not seem to have many globular
clusters, which has perplexed astronomers as most clusters
seem to be located around the central bulge of our galaxy. It
is believed that this cluster and a couple of others may be
impostors, having been brought to the Milky Way by the
Canis Major Dwarf Galaxy, which is in the process of merging with us.

The next-brightest star seen heading further south from
Lepus is Alpha Columbae, which is also the brightest star in
the constellation of Columba. Columba lies between Lepus
and the bright star Canopus in the constellation of Carina to
the south-west, the second-brightest star in the sky. It is easy
to spot and cannot be mistaken as it is rivalled in brightness
only by Sirius to its north. Just a couple of degrees to the
south-east of Alpha Columbae is the orange star Epsilon
Columbae and a line between the two extended by about 5
degrees marks the location of NGC1851, one of the other
globular clusters thought to have been brought to the Milky
Way by the Canis Major Dwarf Galaxy. At 7th magnitude it
is easily picked up with binoculars but appears as nothing
more than a fuzzy star.

To the west of Columba is the constellation Puppis,
which is home to GRB031203, the nearest gamma ray
burster to us, which lasted for just twenty seconds back in
2003. To the east of Columba is the small, faint constellation
of Caelum, which has no stars brighter than magnitude 5,

making it a challenge from all but the darkest sites. Moving further south leads to Pictoris, which looks like a shallow triangle to the south-east of Canopus, with its brightest star, Alpha Pictoris, to its south. About 5 degrees to the east of Canopus is Beta Pictoris, which was the first star discovered to display a protoplanetary disc. These discs ultimately form planets and are evidence that the process of planetary formation is common in the Universe.

One of the showpieces of the sky this month is the Large Magellanic Cloud (LMC), which can be found to the south of Pictoris and looks like a detached part of the Milky Way. As we have seen, it was once thought that the LMC and the Small Magellanic Cloud to its east were galaxies that orbited our own, but it has since been discovered that they are travelling too fast to be gravitationally bound to the Milky Way. In a few billion years' time, they will depart our region of space and slowly fade in brightness. The LMC is 157,000 light years away and covers an area of sky significantly larger than the full moon. On the western edge of the galaxy is one of the largest areas of star formation in our local group of galaxies, the Tarantula Nebula. It can be seen as a 5th magnitude nebula and shows up well with binoculars, although wide-field, low-power telescopes will give the best view.

Epilogue

I CAN REMEMBER WHEN I was just ten years old and my father took me to our local astronomical society. Today, nearly thirty years on, I can vividly recall that evening and my first view of Saturn against the inky black depths of space, which still sends a shiver down my spine. It was a sight that ignited a fire deep inside me and a passion for the Universe which is as strong now as it was all those years ago.

Joining my local astronomical society at that stage in my life was, unbeknown to me at the time, a real turning point and one that carefully nurtured my childlike enthusiasm for all things 'space'. I still have that boyish wonder for the Universe, and these days it is encouraged by a desire to learn about its innermost secrets. It is that knowledge, experience and passion that I hope I have captured and shared with you in this book.

When I was approached to write my guide to the Cosmos after the first series of *Stargazing LIVE*, it was clear that it needed to be different from anything else available. By joining together current theories of the Universe with monthly sky guides even the most novice of sky watchers can grasp the concepts of modern astrophysics. This, along with a new approach to sky guides that makes them relevant to anyone, no matter where they are in the world, means this truly is a book to be enjoyed by everyone.

In the first few pages, I shared with you my Six Top Stargazing Tips and it is these that I would urge you to go back to and look at again, for they will guide you further on your journey of discovery. In particular, I would encourage you to go along to your local astronomical society and join in their events, for the night sky is much more rewarding when shared with friends. Whatever you do and whichever path you take in your continued exploration of the Universe, keep looking up. Keep checking back to my monthly guides so that you can learn your way around the sky gradually. Becoming proficient will take time, so don't expect to remember everything straight away, but stick with it and you will soon be showing others the glorious sights in the night sky.

The relevance of this book to a global audience mirrors the accessibility of the sky, which can be appreciated by us all. Even young children seem captivated by the magic of space and I was very proud to be able to include images taken by schools around the world to illustrate some of the finest objects in the sky. I must thank the team at the Faulkes Telescope Project and Las Cumbres Observatory Global Telescope Network who arranged and helped with this.

I hope you have had fun on this journey through the Universe with me. In my opinion the night sky is one of the most incredible aspects of the natural world and can be easily observed and enjoyed by everyone. Whatever you want to get out of astronomy, either as a casual stargazer or an advanced amateur astronomer, I hope I have been able to ignite a little spark that will continue to grow inside you for many years to come.

Index

THE VOYAGE OF A LIFETIME

A SPACE TRAVELLER'S GUIDE TO THE SOLAR SYSTEM

MARK THOMPSON

'Mark is a wonderful promoter of astronomy, guiding the experienced amateur and complete novice with equal skill and passion.'
Professor BRIAN COX

OUT SPRING 2015
ORDER YOUR COPY NOW